PLANT DISEASES

Their Causes and Control

PLANT DISEASES
Their Causes and Control

Sudhir Chowdhury

BIOTECH
2022
Biotech Books

ISBN 81-7622-175-9

Published by : **BIOTECH BOOKS**
4762-63/23, Ansari Road, Darya Ganj,
New Delhi - 110002
Phones : 011 4300 3222
Email : biotechbooks@yahoo.co.in

PRINTED IN INDIA

PREFACE

This little book is the outcome of much interest in plant diseases shown by the author's students, by his personal friends and by lay visitors to his laboratory. All have been keen to learn something about the nature of disease in plants, the organisms and circumstances which cause plant diseases and the methods of their control and prevention.

In this booklet the causes of plant diseases and the methods of their control have been described in simple language without going much into detail and technicalities. The idea has been to give a general discourse on the subject. It is hoped that students of agriculture, literate cultivators and tea planters who intend to get a preliminary idea of the subject will find this brochure helpful.

The author's thanks are due to a large number of authors whose publications he has freely consulted and wherefrom he has taken much help in writing this booklet. All these publications have been listed in the bibliography.

S. C.

CONTENTS

CHAPTER I

INTRODUCTION

Plant Pathology or *Phytopathology* (Greek: *phyton*, plant; *pathos, disease; logos,* discourse) is that branch of botanical science which treats of the diseases of plants. While there can be no doubt that diseases of plants and in particular of cultivated plants have manifested themselves from time immemorial it is only within comparatively recent years that their study has been taken up in an organised and scientific manner. The application of the scientific method generally to the study of practical agricultural problems is in itself a development which is scarcely older than the nineteenth century, and it is from this combination of science with practice that the modern applied science of plant pathology has arisen.

Plant diseases have been known since the dawn of history. From the earliest times domesticated varieties of plants grown for food or for economic purposes have been subject to the attacks of diseases and early references to the terms 'blight' and 'blasting' show that they obtruded themselves on the mind of the cultivator long before their true nature was known. In Greek literature we find references to plant diseases in the writings of Aristotle and more particularly in those of his pupil Theophrastus, the Father of Botany. The latter describes diseases of a number of cultivated crops such as the various legumes, cereals, fig, olive etc. Among Roman writers the younger Pliny, who flourished in the first century of our era devotes some attention to the subject and makes a number of well founded observations regarding the incidence of certain plant diseases. The ancients had no true conception of the nature and cause of plant diseases and their ideas concerning them were largely of a superstitious nature. The general tendency was to attribute the troubles to God or the gods, as the case might be, or to the influence of the stars and the planets. The Romans, moreover, went a step further and created special gods, such as the God Rubigo to whom they offered up prayers for protection.

The Middle Ages were for plant pathology as for all other sciences, dark ages and we have to come down to the Renaissance, with its general quickening of interest in all intellectual pursuits to find evidence of further progress. The seventeenth and eighteenth centuries witnessed considerable activity in the study of plant diseases. The new studies proceeded in the main from two different angles. The practical gardener or agriculturist was mainly concerned in describing the various diseases to which his crops were subject and in attempting to evolve methods of treatment. On the other hand, the student of human pathology was interested in the diseases of plants from the possible analogies to his own subject which such a study might offer. The whole science of botany had in the old days a distinct medical bias. It is not surprising therefore that the study of plant diseases has been to a large extent modelled upon the older science of human pathology, as it is illustrated for example by the extent to which the methods and terminology of the latter have been freely borrowed by the plant pathologist. It may be remarked here that the two sciences, though in many respects similar, are in others very different.

The elucidation of the real nature of plant disease from the point of view of the causal agent requires in general a considerable degree of optical magnification, and is dependent upon the perfection of the microscope. But even the extended powers of observation which the use of that instrument brought to bear on the subject did not at once lead to a correct appreciation of the true nature of plant disease. Various structures were observed and accurately figured in detail as being associated with the diseased parts, but these structures were interpreted as portions of the plant which had become changed by the disease and not as belonging to a distinct organism which was responsible for the disease. The whole of biological thought was in fact, up to about the middle of the nineteenth century dominated by the belief in the spontaneous generation of life and of disease in particular and throughout the ages down to that time the origin of disease in plants was shrouded in mystery.

The epoch making discovery of Pasteur of the true nature of the process of fermentation and the development of the germ theory of disease mark the true starting point for modern pathological studies, both of animals and plants. The first proof that a specific plant disease was due to the invasion of the tissues of the plant by a definite parasitic organism, was furnished by the German scientist Anton De Bary (1853) who is thus properly regarded as the founder of the modern science of plant pathology at least in its theoretical aspects. De Bary's pioneer researches prepared the way for a host of other workers, at first chiefly in Germany, and later in all civilised countries. The twentieth century has witnessed an extraordinary activity in the study of problems of plant diseases most strikingly perhaps in the United States of America. The economic importance of the subject, as a branch of scientific agriculture, is being increasingly recognised and the prosecution of researches in plant diseases is being more actively pursued by agricultural departments of all countries whose agriculture can claim to be progressive.

CHAPTER II

DEFINITION AND SYMPTOMS OF DISEASE

Disease Defined—Disease in plants may be defined as variation from the normal as expressed either by the checking or by the interruption of physiological activities or by structural changes, which are sufficiently permanent to check development, cause abnormal formations or lead to premature death of a part of the plant or of the entire individual or failure of the plant to produce a commercial product of satisfactory quality and quantity.

Symptoms of Disease—The study of the symptoms is the first step in the examination of disease. By symptoms one means the total of the modifications shown in the plant as the result of disease. In other words the symptoms of disease are those marks or evidences which indicate a diseased condition in the plant. The most common symptoms are listed below. A plant when diseased may show any one, two or more of the symptoms described.

(1) *Change of Colour*—This deviation from the normal may be observed on various plant structures, on the roots, stems and fruits, but the discolourations are very frequent and striking on parts which are normally green, such as herbaceous stems and leaves. Plants, like people, frequently look pale when they are sick. Nutritional disturbances such as lack of iron, excess of lime, excess of alkali, presence of virus, fungal or bacterial parasite, lack of light or low temperature may prevent the production of chlorophyll and cause normal green structures to become pale green, yellowish green, white, red or silvery.

(2) *Leaf Spots*—Discolourations instead of being general or diffuse may show as more or less definite or circumscribed discoloured spots or areas. They form one of the commonest sysmptoms of disease. These spots are very varied in colour, according to the plant and parasite concerned, and also often change colour at different stages in their development. In some cases the colour is not uniform, but zones or bands of different shades may alternate.

Brown spots on Cabbage (After Butler)

Pallid spots are common in the mildews and in the early stages of many other parasitic attacks. *Yellow* spots are caused by rusts and some other fungi. Very often the colour changes to brown or black at a later stage. They are often raised and covered when mature with a yellow dust composed of spores which may be rubbed off on the finger. *Red* spots are found in some rusts and leaf parasites of cereals and sugarcane. In the leaf form of red rust of tea and other plants the spots are raised and often turn a sickly white when old. *Brown* is the characteristic colour of dead tissues and

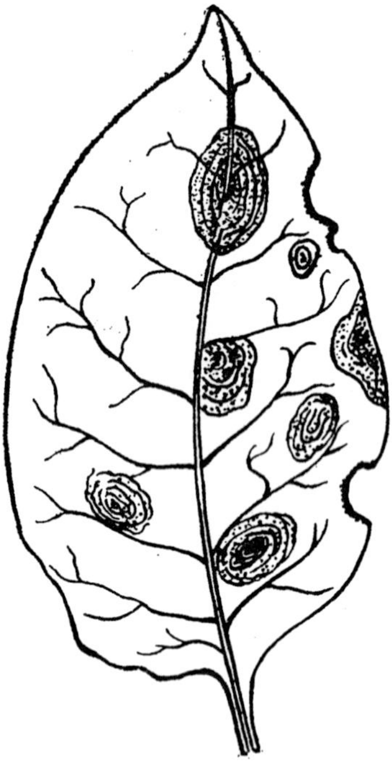

Leaf spot of Potato due to *A. sotani* (After Heald)

is the final colour of the great majority of the leaf spots, although some finally become greyish with age. *Black* spots are almost always an indication of fungus attack. They are found in late stages of rusts and in many *Ascomycetes*, e.g., *Phyllachora, Rhytisma, Diplocarpon. Concentric* bands of colour are found in different shades of brown in spots caused by *Alternaria* and in brown, red and yellow in *Cercospora*.

(3) *Shot Hole*—This name is given to the perforations in the leaves. At first a brown spot appears

but the diseased tissues are soon cut off from the rest of the leaf and ultimately fall off leaving round hole in the leaf. They are common in fruit trees, such as peach and plum and are found in tea and some other

Shot-hole of Peach leaf (After Butler)

plants but rarely in field crops. Shot hole development may result from the attack of bacteria or fungi or from the operation of non-parasitic factors such as frost injury, localised action of toxic chemicals or nutritional disturbances.

(4) *Damping-off*—This name is applied to the sudden collapse of seedlings, which are attacked at the base of the stem and fall over from weakening of the tissues at this point. It results from the rot caused by several fungi e.g., *Pythium, Phytophthora, Rhizoctonia.*

(5) *Wilt*—This name is applied to those cases where a whole plant collapses more or less suddenly and is the result either of the blocking of the water-trans-

porting channels in the stem or the roots or of some injury to the absorbing system which lowers the rate of intake of water below that of the outgo due to transpiration. It may be caused by fungi, bacteria, drought or frost. Some insects or the presence of excess or in-

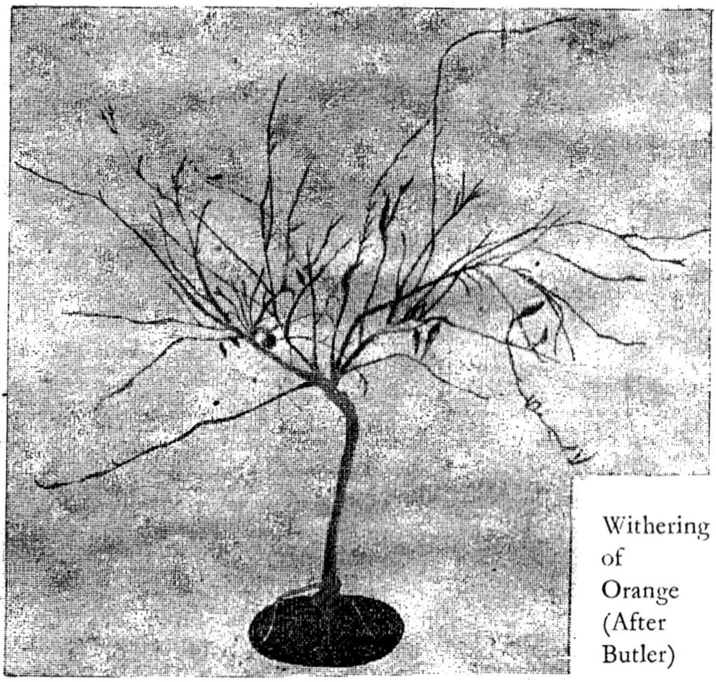

Withering of Orange (After Butler)

jurious salts in the soil may also cause similar symptoms. Drying up of a single branch is also observed in some cases; this is usually due to the infection of the branch by one of the canker or anthracnose fungi.

(6) *Death of parts or Necrosis*—In some diseases death of special parts or organs of the plants, as leaves, stems or twigs, buds, flowers or fruits is the first symptom of disease. The affected structures generally assume the characteristic dark or brown colouration of dead tissue and the accompanying disease is frequently

characterised as a blight. Necrosis may be localised in certain organs or it may be general and result in the death of the entire plant.

(7) *Scab*—This is a condition caused by cracking of the outer layer of fruits or tubers, the broken

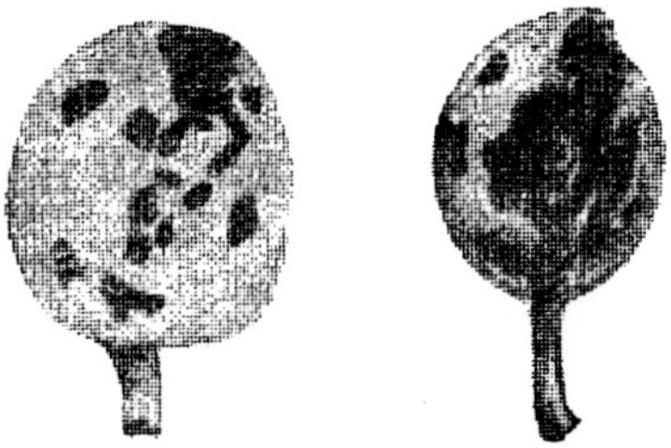

Scab of Apples (After Salmon)

skin becoming dry, flaky, indurate or sometimes corky. The scabs of potatoes, apples and peaches are caused by fungi; citrus scab which is very severe in Assam and prevalent also in other parts of India is due to bacteria. Eelworms may also cause scab of tubers.

(8) *Canker*—Open wounds often of a spreading nature and sometimes surrounded by a raised tumour-like margin, are found on woody stems and even sometimes on annuals like pigeon pea. They are caused usually by parasites that attack the bark and extend as far as the cambium. The raised margins are the results of efforts at repair by the production of callus, which may again be attacked and destroyed before healing has progressed far. Most canker producing fungi are *Ascomycetes* (*Nectria*) but some are rusts (*Peridium*) and *Hymenomycetes* (*Corticium*). Cankers are also caused by insects, frost and bacteria.

(9) *Dwarfing or Atrophy*—Either the entire plant or special organs such as leaves, flowers or fruits may be greatly reduced in size, suggesting such common names as 'little leaf' disease of the apple or 'little peach' with fruits of reduced size. The whole plant

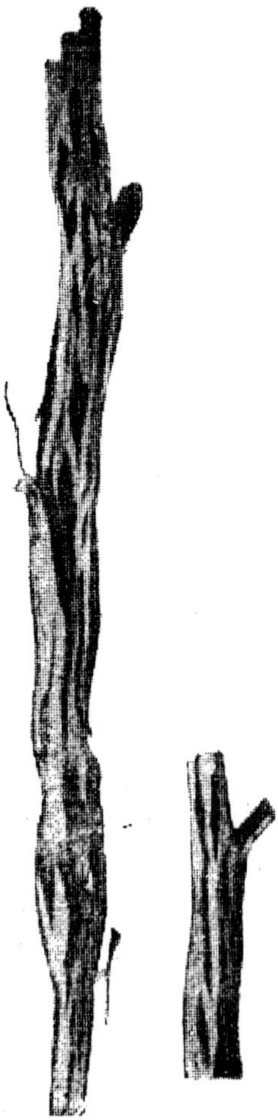

Canker of tea stem (After Butler)

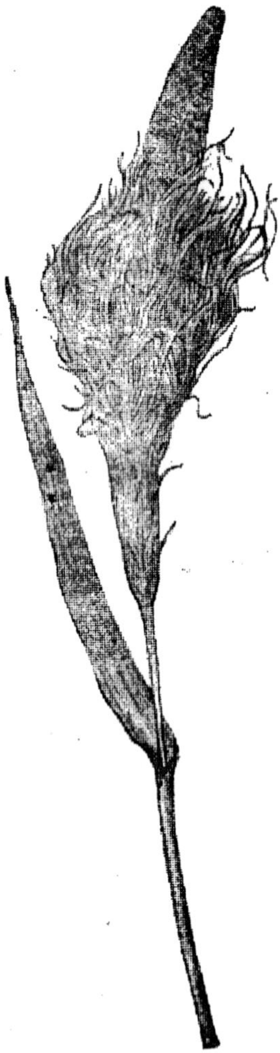

Downy mildew of Bajra, ear changed partly to leafy structure (After Butler)

wheat smut, grains destroyed (After Butler)

may remain stunted from early attacks of mildews, rusts, other fungi and virus. The leaves may be atrophied as a result of the actions of some rusts and species of *Synchytrium;* flowers in poppy attacked by *Peronospora parasitica,* grasses with *Epicloe* and some cereals with *Sclerospora;* fruit in rusted cereals and from smuts that infest the ovary. All these changes can also be caused by insects and several of them by unfavourable climatic conditions, poor soil and the like.

(10) *Increase in size or Hypertrophy*—Abnormal outgrowths of the most varied character are often found on the herbaceous parts of plants and also sometimes on woody stem, and on roots, leaves, fruits and tubers. They may vary from tiny warts, involving only one or a few cells, to the rounded tumours, several inches across, of maize smut. Hypertrophied roots are well illustrated in the club-root or finger-and-toe disease of the cabbage and other crucifers; hypertrophied leaves in the well known leaf curl of peach; hypertrophied flowers or flower parts as in the white rust of the crucifers and hypertrophied fruits as illustrated by plum pockets or bladder plums.

(11) *Transformation of Organs*—This is found chiefly in flowers and results in the change of one kind of floral leaf into another or into ordinary leaves. Stamens may become leafy in the green-ear or downy mildew disease (*Sclerospora graminicola*) of bajra and other grasses; the whole staminate head, the tassel, or the ear may become leafy in the head smut (*Sorosporium reilianum*) of maize; petals may become like sepals, stamens like carpels and carpels leaf like in the white rust (*Albugo candida*) of crucifers. In the ergot of rye and other grasses certain of the ovaries are destroyed by the parasite, and, in the place of the seeds, horny, elongated dark-purple spur-like bodies, the sclerotia or 'ergot' appear.

(12) *Alterations in Symmetry and Habit*—Some

plants which under normal conditions are more or less prostrate or creeping become ascending or even erect when attacked by a fungus parasite. The short unbranched stem with radical leaves of *Launea asplenifolia*

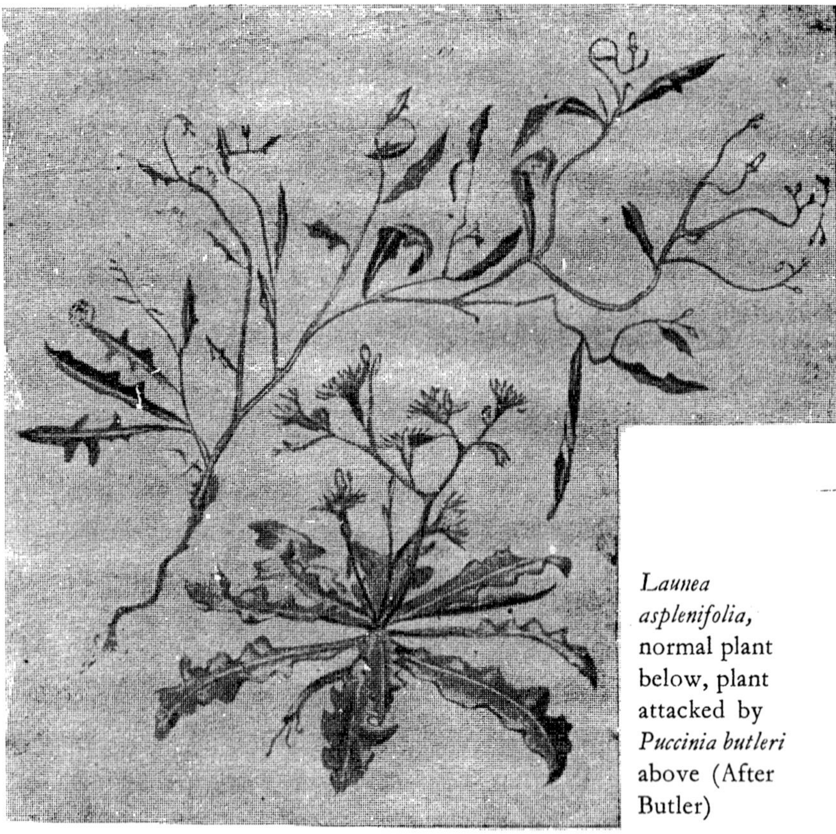

Launea asplenifolia, normal plant below, plant attacked by *Puccinia butleri* above (After Butler)

becomes an elongated, much branched axis with cauline leaves when infected with *Puccinia butleri.* Leaves may be changed from simple to irregularly lobed; single flowers may be altered from regular (actinomorphic) to the irregular (zygomorphic) symmetry and *vice versa.*

(13) *Dropping of leaves, blossoms, fruits and twigs*—This is, of course, to be considered as a symptom of disease only when it occurs prematurely or in excessive amount. The shedding of leaves from the action of parasites may be noted in the leaf spot of groundnut (*Cercospora personatæ*). The shedding of blossoms may be illustrated by the non-parasitic blossom drop of the tomato and grape; non-parasitic dropping of fruits may be illustrated by the shedding of bolls in cotton and the parasitic shedding by the fall of coconuts due to the attack of *Phytophthora*.

(14) *Rot*—Rot is a condition resulting when the cells, wall and content are broken down and more or

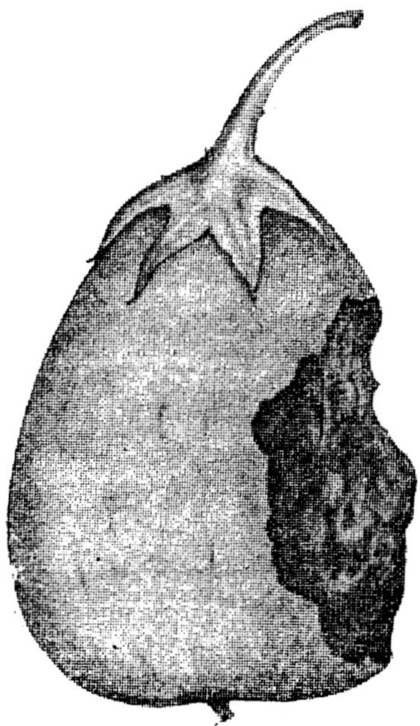

Rot of Brinjal

less consumed by enzymes secreted by the attacking organism. In some cases only the middle lamellae are digested, thus permitting the cells to separate and the tissues to disintegrate. The rotted tissue may be soft or hard, dry or wet and either odourless or possessed of offensive odour. The colour of the rotted tissue also varies. Succulent or woody stems and roots, fleshy leaves, flower buds or fruits may be affected by rot. The character of the rot may depend on the structures involved, the causal factors or complications and external conditions. Rot is often caused by *Phytopthora* and *Pythium* where the green parts are chiefly affected, by such fungi as *Rosellinia* where the roots are concerned, by *Rhizoctonia* where the region affected is usually the 'collar'. Many wet rots are due to secondary saprophytic organisms, chiefly bacteria and moulds, which follow in the track of the parasites.

(15) *Fluxes*—Several trees diseases are characterised by an exudation from the bark of the stem. The nature of the exudation varies in different cases. In the stem bleeding disease of the coconut a colourless or brown, somewhat viscid liquid, rich in sugars and product of decay of the tissues oozes out from cracks in the stem. In rubber canker there is an exudation of *latex* in the older stages. *Resin* is poured out in conifers attacked by various fungi. *Gum* is found on the surface of the diseased parts in the foot-rot of citrus trees.

(16) *Mummification*—The transformation of fruits into shrivelled structures called 'mummies' is a phase of numerous diseases. The formation of mummies is a very characteristic feature of the brown rot of stone or pome fruits. Other typical illustrations may be found in the bitter rot and black rot of apple and grape. 'Mummies' are dried up shrivelled fruits containing the mycelium and sometimes the spores of the parasite: they remain hanging in the tree or fall on the ground.

(17) *Development of dormant, rudimentary or new organs*—The development of dormant or rudimen-

tary structures or of entirely new organs or structures either similar to or entirely unlike any normal parts of the host are symptoms which are associated with certain diseases. Dormant buds may be started into growth, stamens normally rudimentary may grow to full size, extra petals may appear or an entirely dissimilar outgrowth may be formed. New shoots are formed from the base of the rice plant attacked by *Sclerotium oryzae*. The rudimentary stamens in pistillate flowers of *Lychnis dioica* become fully developed when attacked by *Ustilago violacae* except that the pollen is replaced by the spores of the fungus.

(18) *Malformations*—The following abnormal formations may be included under this heading:

Galls or localised enlargements on various organs in the form of small pustules or warts, larger tubercles, tumours or masses of cells making a morbid outgrowth of either fleshy or woody character, in which host

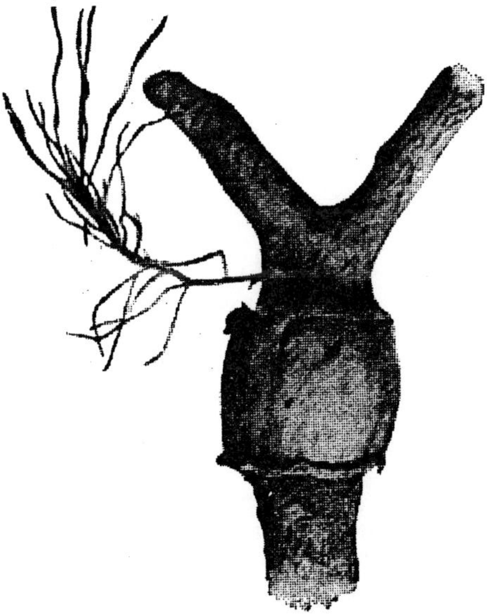

Peridermium galls (After Heald)

tissues and parasites mingle. Galls are overgrowths which result when parasites in the plant stimulate the cells to abnormal activity instead of killing them. A familiar example in this group is the small or large tumours or maize smut.

Intumescences or pustule-like distentions of tissue occurring most abundantly on leaves but also on stems or fruits, due to the abnormal elongation of groups of cells.

Rosettes or closely grouped clusters of leaves caused by the failure of axes to make a normal elongation. This should not be confused with the normal rosette habit of certain plants.

(19) *Witches'-brooms*—This is a condition of closely grouped clusters of fine slender branches, generally arranged more or less parallel to each other and frequently originating from an enlarged axis. It is a dense bushy growth resembling a broom made of twigs. It is caused by parasitic fungi with perennial mycelium in the affected branch. It is chiefly caused by *Exoascaceae* and rusts, common examples being those on barberry bushes caused by *Aecidium montanum* and on deodar caused by *Peridermium cedri.*

(20) *Prolification*—By this is meant the continued development of a part after it has reached the stage

Prolification of Spikelets of *Pennisetum* (After Butler)

at which it normally ceases to grow. It occurs in the ears of bajra and *Setaria* attacked by *Sclerospora graminicola*, the central axis of the flower growing on into a stunted leafy shoot, surrounded at the base by the glumes and stamens.

CHAPTER III

CAUSES OF PLANT DISEASES

The diseases of plants are either due to the attacks of parasites or due to non-parasitic influences. The different causes of the plant diseases may be grouped conveniently as follows:

I. *Parasitic Diseases*—These are caused by (a) plant parasites such as fungi, bacteria, flowering plants; and (b) animal parasites, such as insects and microscopic animals (e.g., eelworms or nematodes). The study of the diseases caused by eelworms comes under the purview of the science of *Phytopathology.* But the study of insects requires a separate field of investigation, *Entomology.*

II. *Non-parasitic Diseases*—They include a great variety of nutritional disturbances due to lack of proper inherent qualities, to improper conditions of soil or air, or to injurious mechanical influences.

III. *Virus Diseases*—Include a group of diseases in which the disturbed condition is the result of an infectious principle, a so-called 'virus' which can be transmitted from diseased to healthy plants and can communicate the disease.

I. Parasitic Diseases

(a) *Fungi*

The most widely distributed and most destructive of plant diseases are caused by organisms known as fungi. Fungi are an enormous assemblage of organisms which are usually regarded as plants but which differ from plants in general in that they do not possess chlorophyll. They vary in size from the large mushrooms and toadstools to forms which are invisible to the naked eye; there are species which are edible and others which are amongst the deadliest of poisons; some species are parasites of plants and animals whilst others inhabit the soil, manure piles, plant debris and a host of other substrata. The fungi which are the concern of Plant Pathology are those which cause plant diseases.

Mycelium

The vegetative body of the fungus is called the *mycelium*. It consists of a delicate, much-branched, tubular or filamentous structure. Each individual

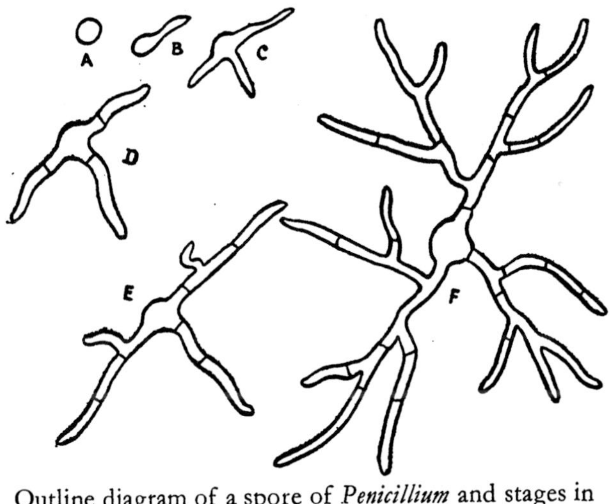

Outline diagram of a spore of *Penicillium* and stages in its germination to form hyphæ and a young septate mycelium (After Heald)

filament is called a *hypha*. The hypha may be septate or non-septate; when septate the constituent cells are either uninucleate or multinucleate, when non-septate the nuclei are distributed irregularly in the protoplasm.

In most cases the hyphae are richly branched; they elongate by apical growth and as a rule spread loosely through the substratum; in some cases they may be densely interwoven or even compacted into solid bodies, called *sclerotia.* The fungus body of some forms, either primitive or degenerate types, may consist of only a single globular cell, no longer than individual cells of a hypha.

Reproduction

Fungi commonly reproduce by *spores* which function as seeds, serving the same purpose for fungi as the true seeds do for seed plants. The spores are very small bodies of microscopic size which are either cut off from hyphae or produced by specialised structures known as *sporophores* or spore fruits. They are cells set aside for reproduction with their protoplasmic content enclosed in a membrane; they may be smooth or scluptured, hyaline or coloured and of various shapes and forms. In a single fungus family as well as in a fungus genus various kinds of spores may be formed, but in so far as is known, all are functionally alike in that each kind of spore can develop a plant similar to the mother plant in all its developmental cycle.

Spores may be either *sexual* or *asexual.* When spores are formed as the result of a breeding act, that is the union of two distinct and separate cells or elements which represent male and female they are called *sexual* spores; when they are formed direct from hyphae without the intervention of a breeding act they are called *asexual* spores.

Kinds of Spores

Various kinds of spores are produced by fungi. The following are the most important types of spores: (1) *Chlamydospores,* thick-walled resting spores, formed by thickening and swelling of certain cells of a hypha or of an entire hypha: (2) *Zoospores* or *swarm spores,* naked protoplasmic masses of varying form provided with one or more delicate vibratile

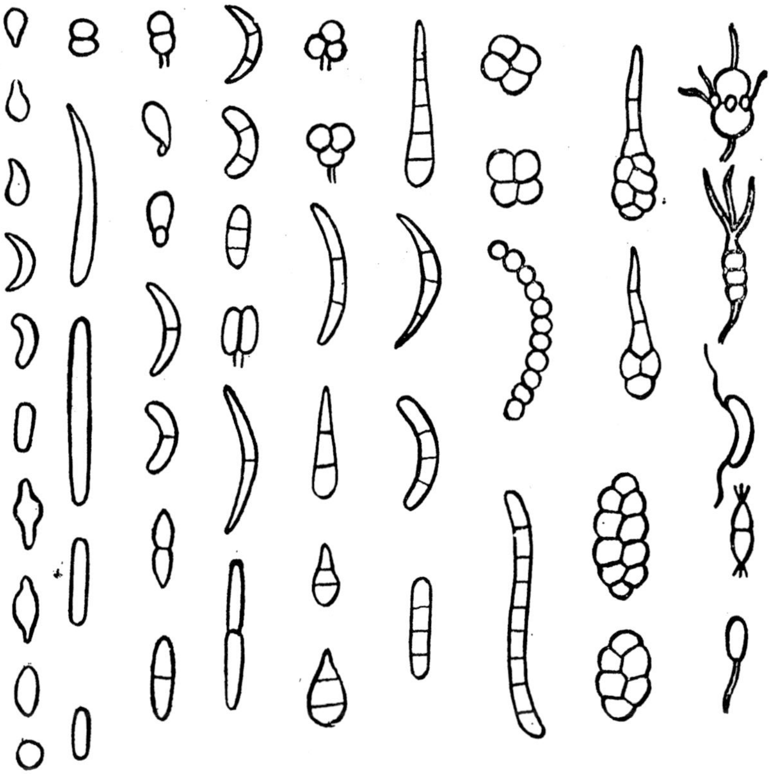

Diagrams of various forms of spores (After Heald)

filaments, the *cilia,* by means of which they are able to swim about; (3) *Conidia,* spores of various sizes, forms, septation and colour, formed by budding or abscission from special hyphae (*conidiophores*); (4) *Oidia,* in some fungi portions of the mycelium form close series of very short cells, which may remain roughly rectangular or may round up to resemble a chain or beads. The mycelium then breaks up at the septa and the fragments, known as *oidia,* function as spores; (5) *Ascospores,* typically eight in number developed within a saclike or club-shaped structure, the *ascus;* (6) *Basidiospores,* developed on the tips of either two or four (rarely more) slender terminal projections, or *sterigmata,* from a globular, cylindric or club-shaped cell, the *basidium;* (7) *Zygospores,* formed by the union or conjugation of two equal and similar cells or gametes; (8) *Oospores,* formed by the union or fertilization of a large passive, female gamete with a small active male cell or gamete. Besides there are other kinds of spores but the most important ones are the *uredospores, teuletospores, aeciospores* and *pycniospores* which are exclusively found in the rust fungi.

Sporophores

Spores are produced by specialised structures known as sporophores or spore fruits. They are of numerous kinds. The most important ones are mentioned below: (1) *Conidiophore,* a specialised hyphae bearing conidia either at the tip or more rarely, along its length. In some genera the conidiophore terminates in a series of branches of more or less definite form, on which the conidia are borne; (2) *Coremium,* an erect compact cluster of conidiophores grouped together to form a sterile stalk and a terminal fertile or spore-bearing head; (3) *Sporangium,* a closed receptacle, usually round or pear-shaped, borne on a stalk (the *sporangiophore*) and containing an indefinite number of spores, either non-motile or motile by cilia; (4) *Pycnidia,* fruit-bodies which are typically flask-shaped, but are often irregular, generally embedded in the substratum and opening on the surface by a pore, or *ostiole,* through which the spores, *pycnospores,* are extruded; (5) *Sporodochium,* a cushion-shaped, tightly packed cluster of short conidiohpores;

(6) *Acervulus*, typically a saucer-shaped, depressed structure which bears conidiophores over its exposed surface and sets the spores free by the rupture of the superficial cell layers of the host; (7) *Sorus*, a small or

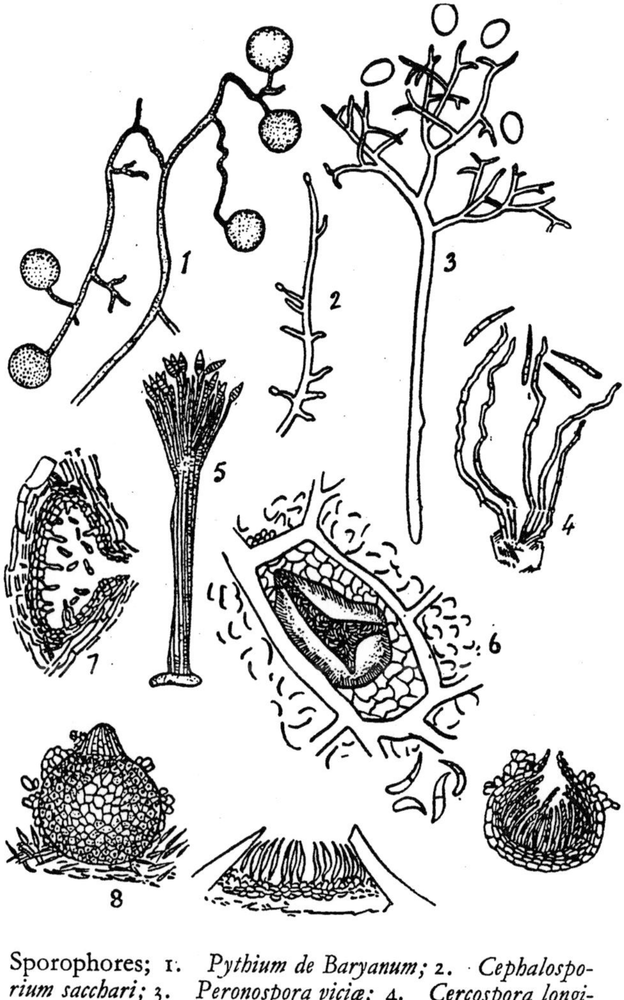

Sporophores; 1. *Pythium de Baryanum;* 2. *Cephalosporium sacchari;* 3. *Peronospora viciæ;* 4. *Cercospora longipes;* 5. *Arthrobotryum* Sp. 6. *Marssonia juglandis;* 7. *Ascochyta Pisi;* 8. *Nectria Bolbophy lli.* (After Butler)

large mass or cluster of spores borne on short stalks; (8) *Pycnium* or *Spermogonium*, a small, flask-shaped body which produces minute, bacteria-like bodies, the *pycniospores* or *spermatia;* (9) *Aecium*, a cluster of closely packed, yellow or orange-coloured spores, *aeciospores* either naked or in cup-like receptacles; (10) *Ascocarp*, consists of a group of asci, protected by a common wall of sterile filaments known as the *peridium*. The young ascocarp is globose and usually completely closed; it may retain this form at maturity, opening only by the decay or irregular splitting of its walls; it may assume a flask-shaped outline, opening by a terminal pore, the *ostiole*, or it may spread out to form a cup in the concavity of which the asci are fully exposed. The cup-shaped ascocarp is known as an *apothecium*, to the other forms the term *perithecium* is applied.

How Fungi Live

Since fungi never possess chlorophyll they are powerless, as a rule, to prepare food for themselves from the simple elements of the air, water and soil. They require their food, therefore, ready prepared. According to the manner in which fungi obtain their organic food they are divided into two great classes: *Saprophytes* and *Parasites*.

Saprophytes obtain their organic food from the dead tissues of animals or plants or from substances derived from them. Hence they are found in large numbers in the soil, in manure piles and in rotting leaves, twigs, wood and other plant and animal refuse. They grow as moulds on bread, jam, fruits and on boots, furnitures and backs of books during the rains. In forest they are found in every bit of old timber and in water on submerged organic materials.

Parasites obtain their organic food from the living tissues of animals and plants. Hence they live with consequences to the host that vary from trifling inconvenience to complete destruction. The great majority of the parasitic fungi feed on living plants useful to man. Some are also found to parasitize man, domestic animals, fish and insects. The few cases of beneficial parasites from the point of view of man are those which attack noxious insects and occasionally destroy large number of them.

Several different classes of parasites may be distinguished. *True* or *Obligate parasites,* for instance, pass through the whole of their life cycle on living plants and cannot be grown on dead or artificial food material and the death of the host involves the death of the fungus. *Hemi* or *Partial parasites* are those that feed usually on living tissues but can, at need, pass through a part of their lives as saprophytes. *Hemi*-or *Partial-saprophytes*, on the other hand, are those which usually grow on dead or decaying matter and are capable of passing through the whole of their development as saprophytes, but have at the same time the faculty of attacking living tissues under certain conditions.

Ecto- and *Endo-Parasites*

The parasitic fungi may be divided according to their place of occurrence and their mode of attack on the host into two categories which may be designated as *ectoparasites* and *endoparasites.*

Ectoparasites live on the surface of the host and obtain their food through the outer cell walls without deeply penetrating into the tissues. They cover the surface of the leaves, stems and other parts of the affected plant with a web-like mycelium, which may be lifted off, leaving the surface bare and the walls of the outer cells intact. They feed usually by means of special out-growths from the outer mycelia known as *suckers* or *haustoria.* These arise from hyphae in contact with the epidermis and penetrate wholly or partly through the outer epidermis; the food is absorbed from the deeper layers of the plant by these haustoria which in turn nourish the outer mycelia. In several cases no haustoria have been detected; in such cases it is presumed that the food is absorbed directly from the outer cell across the unbroken cell wall.

Endoparasites, on the other hand, penetrate into the plant and develop the vegetative mycelium within the tissues of the plant. They obtain their food from the cell sap of the living plant cells, either by the hyphae or haustoria from them growing directly into the cells in search of it or by absorption through the walls in the case of such fungi as live between and not inside the attacked cells and which have no haustoria.

Methods of Host Penetration

When the spore of a parasitic fungus falls on the leaf or any other part of a host it awaits favourable conditions for further development. When the moisture, temperature and other conditions are favourable, the spore *germinates,* that is sends out the germ tube. The germ tube enters the host plant either directly through the stoma or by direct passage through the outer wall of the epidermis which it dissolves by secreting enzymes. In woody or corky parts, the germ tube may enter through the air passage of the lenticel, through cracks in the bark, or directly through the cell walls. Most of the stem parasites of trees enter through wounds especially those due to the breaking of branches or the action of insects; in some cases they can enter through dead branches.

Infection of roots may be by direct boring through the uninjured epidermal or cortical cells or through wounds or cracks in the bark. Usually the first type occurs in young root, the second in the old ones.

There are parasites which can only enter through definite parts of the host plant, though their subsequent growth may be diffused and they may form their spores at a considerable distance from the point of entry. In oat smut, for instance, infection occurs through the mesocotyl of the germinating seed, but the spores are formed some months later in the ears. In the loose smuts of wheat and barley and in some *Sclerotinia* infection occurs through the stigma of the flower; in ergot infection occurs through the young ovary.

Absorption of Food

Fungi, both parasitic and saprophytic, can absorb food from the substance on which they are growing. But very often the cell substances are present in an unsuitable and unavailable form and they must be changed to enable them to serve as food for fungi. Starch, for instance, has to be changed to sugars, proteins to peptones, glucosides to glucose, fats to fatty acids and glycerine. Such changes are brought about by enzymes in which fungi, both parasitic and saprophytic, are exceedingly rich.

Dissemination of Fungi

A fungus may be disseminated either by its spores or by its mycelium. The great majority of fungi are, however, ordinarily disseminated by spores. When a fungus is disseminated by its mycelium, it is usually carried with the substance on which it is growing. Several fungi are distributed by their mycelia; the well-known potato blight fungus, *Phytophthora infestans*, may be cited as an instance.

The dissemination of fungi is carried out by a number of agencies. Of all the agents concerned wind ranks the first. Fungi produce very large number of spores which are very minute bodies and very light in weight. They are, therefore, easily taken up into the air as impalpable dust and carried from place to place.

Many spores are again produced in receptacles which are sometimes provided with arrangements for expelling the spores forcibly into the air after which they may be carried by air currents.

Water may also serve for the dissemination of fungi. The water moulds, the chytrids, the white rusts and some of the downy mildews produce actively motile reproductive bodies. Such forms depend wholly or in part upon liquid moisture as a medium in which they may develop and through which they may migrate by the action of their flagella or cilia.

Flood water, irrigation water or steam flow carry many fungi from one place to another. Water can carry a load of spores from one field to another and increase the chances of infection. Sclerotia of *Sclerotium oryzae* and *S. rolfsii* are carried from one field to another by flood and irrigation water. Sclerotia of ergot affecting aquatic grasses are carried to new hosts by stream water.

The dripping of moisture from heavy dews or the run off from rains may transport non-motile spores: those which have been lodged on plant parts by wind or other agencies and those which are residual in origin but are held together by a mucilaginous matrix in which they are embedded.

Insects may act simply as carriers of infective material adhering to their bodies and in certain cases make the inoculation by feeding wounds, either by chewing or sucking mouth parts. The other forms of animal life credited with the dissemination of diseases are nematodes, slugs or snails, birds and wild or domestic animals.

Man is an important agent in fungi distribution. In all of his commercial transactions, such as shipments of grains, introduction of plants and moving of commodities fungi of many kinds may be introduced and spread over vast areas.

True seed and also vegetative reproductive structures such as tubers, rhizomes, cuttings and grafts may carry the fungi (mycelium and spores) on or within their bodies or as separate bodies mingled with the seed. Some of the recorded cases of surface-borne fungi on seed are bunt of wheat, covered smut of oats and barley and juar smut. The fungi which cause the anthracnose of wheat, the spot blotch of barley and loose smuts of wheat and barley are carried within the seed.

By the transport of soil, litter, compost or manure fungi may be introduced into new environments. Fungi may also be spread by the ordinary operations of transplanting seedlings, by watering, by spraying, by picking or harvesting operations and by threshing and ginning. Infected soils may reach other fields in various ways as on the boots or feet of farm workers and pedestrians or on agricultural implements.

Life History of Parasitic Fungi

One of the most important question to be settled with parasitic fungi is—where and how do the parasitic fungi live during the absence of their normal host plants? In the plains of Assam, for instance, the potato season lasts from October to March. After this period no potato plant is found any where between April and October. The fungi which are parasitic on potato have, therefore, to pass through a period of about seven months in some other way than as active potato parasites. The same is true with other fungal parasites which do not live on the stems and roots of perennials or on the leaves of the evergreens. Various methods are adopted by parasitic fungi to tide over this difficulty; some of these are very interesting though they are not in all cases by any means completely understood. Some of the more completely known methods of their perpetuation are listed below:

(i) *Perpetuation by means of Spores*—Many parasitic fungi form spores which are very resistant to adverse circumstances and remain viable for a long period, even for years. Such spores help to an enormous extent the perpetuation of many fungi. These spores either live in the soil, plant debris, manure pile or get mixed with the seed during harvest, thrashing and storing operations and remain in storage with the seed. The spores of *Erysiphe graminis, Plasmopara viticola, Venturia inaequalis, Alternaria sp., Fusarium sp., Gibberella sp., Tilletia tritici, Cercospora beticola, Glomerella gossypi* and many other parasitic fungi survive in the soil or in the plant debris left in the soil during the absence of their respective host plants and infect and attack the hosts when they are again sown. The spores of the fungi causing the covered smut of barley and the oat smuts, on the other hand, remain clinging to the surface of the seeds and lie in storage along with the seeds. When the seeds are sown the spores of the fungi are also sown along with the seeds. The spores of the fungi germinate along with the seeds of barley and oats and attack their respective hosts.

(ii) *Perpetuation by Dormant Mycelium*—There are numerous instances in which the parasitic fungi are carried internally in seeds and other propagative parts in the form of dorment mycelium. The loose smuts of wheat and barley are among the most frequently cited examples of this type. The mycelia of these fungi remain in the seed in a dormant condition until the seeds are sown, when the fungi resume activity and pass up into the seedlings, keeping pace with the growth of the hosts until they are able to form new spores in the young ears. Bean anthracnose (*Colletotrichum lindemuthianum*) is another good example of a fungus which perpetuates as mycelium within the tissues of the seed. *Phytophthora infestans, Verticil-*

lium albo-atrum and *Fusarium sp.*, are perpetuated in the form of dormant mycelium in potato tubers.

(iii) *Perpetuation by Perennial Mycelium*—Fungal parasites of many perennial plants persists in the perennial parts of their hosts and pass out into the new shoots or leaves arising from the infected parts. The above ground parts of the weed *Launea asplenifolia,* which is attacked in winter by a rust, dies down at the beginning of the hot weather. The weed has a perennial root-stock which survives underground. The mycelium of the rust fungus perenates in this root-stock during the hot months and attacks the new growths of the weed which come out during the beginning of the winter. Several other similar cases are known amongst rusts, smuts and other parasitic fungi. The cherry witches'-broom fungus (*Exoascus cerasi*) penetrates the tissues of the branch of the host and lives there perenially without killing the twig. Several members of the *Erysiphaceae* are known to persist in a dormant mycelial condition on perennial parts of the host.

(iv) *Perpetuation by Sclerotia*—Many parasitic fungi form thick mycelial bodies, usually called *sclerotia.* These are thick-walled compact bodies, resistant to extremes of cold and heat and long-lived. They remain viable in the soil, manure or plant debris for a long time and infect the host plant when it becomes available. *Sclerotium oryzae* and *S. rolfsii,* for example, live between successive crops of their normal hosts chiefly as sclerotia in the soil. *Rhizoctonia solani* may also be perpetuated through sclerotia surviving in the soil but it is also known to be carried through its sclerotia borne externally on the surface of the seed tubers of potato. The ergot of grains and grasses is perpetuated through the sclerotia of the fungi.

(v) *Perpetuation on other or Alternate Hosts*—In many cases the fungal parasite is capable of living on more than one host plant, sometimes on a large number belonging to several different natural orders. This naturally helps to keep it alive when suitable food plants are found throughout the year. *R. solani*, for instance, has a wide host range and some of its hosts are found growing in our fields all throughout the year. Some of the rusts which are heteroecious have alternate hosts and they survive on their second host during the absence of the first and *vice verse.* Some fungi are also known to live on weeds, grasses and other plants during the absence of their normal hosts.

Finally, there are cases in which we do not know yet clearly how persistence of the parasite is effected.

Heteroecism

The great majority of parasitic fungi pass through their life history in or on the tissues of a single host plant; such forms are said to be *autoecious* in contradistinction to the *heteroecious* forms which require two alternating hosts for the completion of their development. With the exception of the ascomycetes species *Sclerotinia ledi* all known heteroecious fungi belong to the rusts.

Polymorphism

One of the most curious phenomena in the life history of parasitic fungi is their capacity for assuming different forms at different periods of their full development. To this the name '*polymorphism*' is given. The Ascomycetes and the rusts are particularly rich in polymorphic species. Polymorphisms, however, occur to some extent in all groups. The example of the common pea rust, *Uromyces fabae*, may be cited to illustrate this. The fungus first makes its appearance as yellow dots in clusters on the stem and leaves of the pea plant. On microscopic examination they are found to consist of flask-shaped fruit bodies containing large numbers of minute spores. This spore fruit of the fungus is called a *spermagonium* and the spores produced inside it are called *spermatia.* Very soon, larger yellow eruptions appear in the same parts of the stem and leaves; these are known as *aecidia* and they bear chains of rather large, yellow spores, known as *aecidiospores.* A little later, pale brown pustules, the *uredo sori* appear scattered all over the plant and bear the characteristic *uredospores* of the fungus. Still later, large, blackish-brown pustules are found chiefly on the stems. These are the *teleuto sori,* in structure similar to the uredo sori but bearing the *teleutospores,* at the tip of numerous stalks arising from the base of the sorus.

Specialisation of Parasitism

Fungi vary in the extent to which they are adapted or restricted to a particular habitat; in some species the range is wide, in others the range is narrowed. Many fungi are again limited, either as parasites or saprophytes, to the members of a particular family, a particular genus, even a particular species.

Amongst those fungi which are capable of attacking several different species of plants, many have developed into distinct races each of which though outwardly similar to the other, is restricted to one, or a few only, of the host plants. A single species of fungus such as *Puccinia graminis* may include a number of these races, all quite similar in structure and not to be distinguished from one another in any other way than by their capacity for living on certain hosts. The difference between them is physiological, not botanical. The name *physiological* or *biological* species or forms has been given to these races. To this splitting of a parasite into specialised races on different host plants, the term *specialisation of parasitism* is applied.

Classification of Fungi

The fungi as a whole are divided into three main classes according to the septation of their mycelium and the character of their principal spores: *Phycomycetes, Ascomycetes* and *Basidiomycetes*; and a fourth class, whose perfect stage is unknown, the *Deuteromycetes* or *Fungi Imperfecti.* The fungi are further sub-divided into groups, alliances and families.

Phycomycetes—This lowest class of true fungi is often known as the alga-like fungi, because they resemble some groups of the green alga. Many of the aquatic fungi and simpler land forms belong to this class. The vegetative hyphae do not unite in strands or form a tissue and usually remain without cross partitions or septa except at the base of fruiting structures. The sexual reproduction is of a distinct and simple type, and the nuclei usually fuse in the cells in which they come together. Most of the species also have an asexual stage, which is usually more conspicuous than the other. The class is divided into three sub-classes, *Archimycetes, Oomycetes* and *Zygomycetes.*

Ascomycetes—The vegetative hyphae have numerous cross partitions or septa and the sexual spores are produced in a mother cell known as the ascus, arising from special cells. The asci are usually borne within or on a special body the ascocarp, formed by the union of a number of vegetative cells. The sexual reproduction is more obscure than in the Phycomycetes. A hollow sphere or flask-like body enclosing the asci is known as a *perithecium*, but a wide open receptacle is known as an *apothecium*. These special bodies may in some forms be within or on a tissue-like structure formed from the coalescing and compacting of hyphae and known as a stroma. In addition to sexual spores there may be asexual forms of the most diverse sorts. The class is divided into three sub-classes, *Plectomy-*

cetes, Discomycetes and *Pyrenomycetes*.

Basidiomycetes—This third main class of the true fungi includes a large number of known higher fungi, most of the so-called mushrooms or toadstool forms, as well as the smut and rust fungi. They are very diverse in size and form but include many of the larger definite forms whose sporophores may be readily seen without the aid of the microscope. The Basidiomycetes have only a very reduced form of sexual reproduction. Special spore-bearing cells (*basidia*), usually club-shaped, arise, from which there project two or four minute short stalks (*sterigmata*), each bearing a one-celled spore (*basidiospore*). This class is divided into three sub-classes: *Hemibasidiomycetes*, or smut fungi; *Protobasidiomycetes*, which include the rust fungi; and *Autobasidiomycetes*, which include the mushrooms, puffballs, and most of the larger fungi.

Deuteromycetes—This is largely an artificially arranged class of fungi appended for convenience in order to classify fungi either that have lost some of the characters that would place them in the other three natural classes or whose life histories are yet too imperfectly known to enable them to be placed elsewhere. A large number of them are probably imperfect stages of Ascomycetes, fewer are stages of the Basidiomycetes, and possibly a very few are imperfect stages of Phycomycetes. Those classified here are devoid of known sexual reproduction or sporangium formation. The reproduction is usually by means of conidial spores. Three sub-classes have been formed based on the manner in which the spores are formed, or the manner in which they are arranged on the hyphae or within special receptacles: *Sphaeropsidales*, *Melanconiales* and *Moniliales*.

(b) *Bacteria*

One does not usually associate bacteria with

diseases in plants, but they are, nevertheless, frequently agents of such disease. In recent years many diseases of bacterial origin have been discovered and described. Man's chief interest in bacteria is usually centered in the diseases of man which are so largely caused by bacterial action. But in 1896 Smith stated that 'there are in all probability as many bacterial diseases of plants as of animals.' In 1920 the same author reported the occurrence of specific diseases on host scattered through more than 150 genera and over 60 families. Many of the hosts are important cultivated plants and wild plants of economic importance. Each year since the publication of the statement quoted has witnessed numerous additions to the long list of recognised bacterial diseases of plants.

The proof that the bacteria are the cause of plant disease (fire blight of the pear and apple) was first furnished by Thomas Burrill, Professor of Botany in the University of Illinois (1879-1881).

Forms of Bacteria

The body of a bacterium consists of a single cell into the minute structure of which it is not necessary to enter here. It is sufficient to note that there is a central body of protoplasm which stains readily with aniline dyes of various kinds, and a skin or capsule, usually so thin as to be practically invisible except by the aid of the highest powers of the microscope. The capsule or cell wall is not composed of cellulose as in the higher plants, but of a material apparently allied to the protein substances.

The whole body or cell is met with in many forms. Some kind of bacteria are spherical, each individual being then termed a *coccus*. When much longer than broad the bacterium is known as a *bacillus*; a comma-shaped form is spoken of sometimes as a *vibrio*; a rigid spiral form in shape like a cork-screw is a *spirillum*.

Under certain circumstances bacterial cells may become distorted and irregular in shape. Such are known as involution forms.

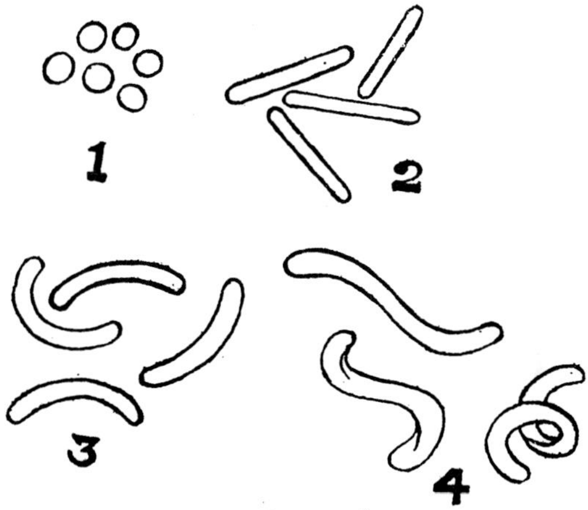

Forms of Bacteria

1. *Coccus*, 2. *Bacillus*, 3. *Vibrio*, 4. *Spirillum*

Size of Bacteria

Bacteria are amongst the smallest of all living things. Their dimensions are usually stated in thousandths of a millimetre. One thousandth of a millimetre (.001 mm.), known as a micron, is denoted by the sign μ and is equal to about 1/25000 of an inch. Cocci range from less than 0·5 μ up to 2 μ in diameter; bacilli from 0·5 μ or less in breadth to 10 μ in length; the majority, however, are much shorter than the latter figure. On account of their minute size they are, when dry, readily blown about by the slightest breeze, and find their way into any crevices wherever dust can penetrate. They are carried in streams of water, and are to be found in and about all foods, milk and other liquids, in houses and stables and in the soil, on the skin and in the stomach and intestines of all animals. They are, however, never met with in the blood or tissues of healthy animals or plants and are rare in water which has come from deep wells or springs or has been filtered through great depths of soil and rocks.

Movement of Bacteria

When bacteria are viewed under the microscope in a living condition many of them are seen to move. They can rove from one part of a drop of water to another as readily as fish in a pond. This is brought about by the lashing action of one or more exceedingly delicate hair-like *cilia* or *flagella* which are attached to the outer coat or capsule of the bacterium. Some organisms possess only one flagellum at one end (*monotrichic*), others two or more in a tuft at one or both ends (*lophotrichic*), while many bacilli have many flagella distributed all over the cell wall (*peritrichic*).

Reproduction in Bacteria

Reproduction among the bacteria is largely asexual and takes place ordinarily by what is known as binary fission. In addition to this a number of bacteria go into a resting stage or produce spores. The spore formation is not, however, a method of multiplication because usually only a single spore is formed in a cell, but serves to tide the organism through unfavourable conditions.

(i) *Vegetative Reproduction*—When bacterial cells reach their adult normal form, if the conditions of temperature and nutrition are satisfactory, a partition or septum is developed across the cell which divides it into halves. The two halves or daughter cells, then grow to an adult size and undergo similar division. The process is repeated and vast numbers may arise in this way. A bacillus will divide into two bacilli in about twenty minutes; each of these new organisms will at once begin to grow, and in twenty minutes division will again take place. This means that under ideal conditions a single bacillus would in eight hours produce 16,000,000 descendants. The

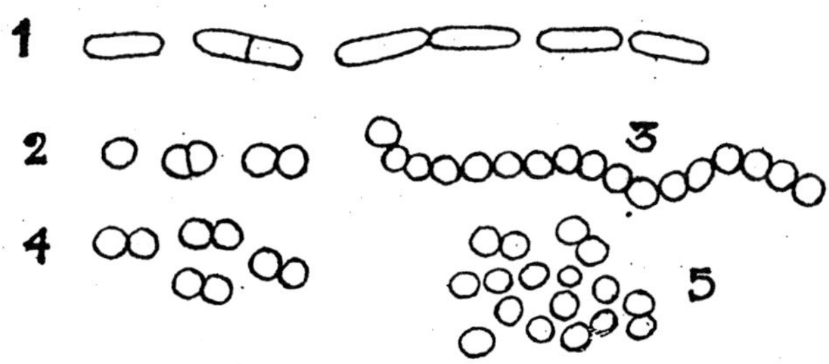

Vegetative Reproduction of Bacteria, 1. *Bacillus* with successive divisions, 2. *Coccus* giving rise to chain, 3. Pairs, 4. Irregular groups

weight of a single organism is infinitesimal but in twenty four hours it would, if allowed complete growth, produce about 500 tons of bacilli.

It follows that if ideal conditions were available to bacteria the world would be swamped by their growth. There are, however, many factors operating against this possibility. Only the common organisms grow at the rate mentioned and many require days or even weeks to produce any appreciable weight of material.

(ii) *Spore Formation*—A considerable number of bacteria form spores within the cell. Because they are formed within the cell, they are spoken of as 'endospores'. Endospores are formed by the bacilli and the spirilla and their chief value to the cell is their ability to resist unfavourable conditions which to the ordinary vegetative form of the cell would prove disastrous. The spore is formed by the contraction of the protoplasm within the cell. The protoplasm of the spore is denser than that of the original mother cell and the coat of the spore is a somewhat thick membrane. The number of spores which originate in each bacterium is generally one, though in one or two kinds two or more spores are seen in each cell.

When placed under favourable conditions of temperature, and provided with a suitable moist nutrient and a proper air supply the spore germinates by the rupture of the spore wall.

Food of Bacteria

Like the fungi bacteria are also plants devoid of the green colouring matter called chlorophyll. Hence, they cannot manufacture their own food materials. In respect of the source from which the organic compounds are derived two classes of bacteria may be recognised, viz., *saprophytes* which obtain their food from the dead and decaying bodies of plants and animals, or the lifeless organic matter in the soil, milk, water and of other common materials; the *parasites* which feed upon materials present in the tissues of living organisms, plants and animals.

How Bacteria Invade Plants

The external surface of the plant body of seed plants is covered in large part by an epidermis with an external cuticle or also with cuticularised external walls or by a more impervious layer or cork cells, the periderm. Some parasitic fungi are able to make their way through such unbroken epidermal walls, but pathogenic bacteria seem unable to penetrate cuticularised walls or layers of cork cell. This leaves wounds or mechanical injuries of various kinds, stomata, water pores, nectaries and lenticels as the possible avenues of entrance.

Location of Barteria in Diseased Tissue

The majority of bacteria which cause disease in plants enter into the tissues of various organs or parts by some of the methods stated above but in a few cases diseases are caused by the development of the pathogene between certain closely oppressed organs, between the glumes of cereals and grasses and between the outer petals of unopened buds.

Bacteria which actually penetrate the tissues may be: (a) *intercellular* or in the spaces between the cells; (b) *inter-vascular* or in the water-conducting vessels of the xylem; (c) *intra-cellular* or within the interior of cells.

Action of Bacteria on Plants

There may be a mechanical splitting, tearing or crushing due to the enormous multiplication of the bacteria within the confined spaces but the principal actions are of a chemical nature. Some of the important changes are: (a) the separation of cells from each other by the digestive action of enzyme upon the pectic substances of the middle lamellae, leading to the production of cavities in various tissues; (b) the production of enzymes which convert starches into sugars, complex sugars into simpler forms or digest and make possible the assimilation of proteids and other nitrogenous compounds; (c) the formation of injurious acids, alkalis or toxic substances which either inhibit physiological processes or actually kill the protoplasm by their poisonous action; (d) the production of substances which stimulate cells to abnormal activity.

Dissemination of Bacterial Diseases

Bacterial diseases may be introduced into new environment or spread from diseased to healthy plants in a great variety of ways. Diseased plants may harbour the disease and serve as centers of infection when transported to distant points or when allowed to remain in fields or orchards. Some of the more important agents of transmission are: (a) seed, fruits, bulbs, tubers, grafts, cuttings and other propagating stock; (b) insects and other animal life including birds, mollusks and worms; (c) contaminated composts and manures.

(c) *Parasitic Flowering Plants*

Besides the diseases caused by fungi and bacteria there are a few flowering plants which are parasitic on other living plants. In the great majority of the cases the damage done is slight or the host plant is of little economic importance but there are a few instances in which they invade valuable crop plants and cause considerable monetary loss.

Some of these are parasitic on the roots of the host plant while others grow attached to the stem of the supporting plant. Some are devoid of chlorophyll and entirely dependent upon the host plant while others have more or less chlorophyll and are only partly dependent on other plants for foods or raw material.

The common flowering plant parasites may be conveniently grouped as follows:

(i) *Stem Parasites*:
- (a) Entirely dependent—Cuscuta
- (b) Partly dependent—Mistletoe

(ii) *Root Parasites*:
- (a) Entirely dependent—Broomrape
- (b) Partly dependent—Striga

Cuscuta

Cuscuta (*Swarnalata*) is a slender twining vine with minute, rudimentary scale-like leaves and almost or entirely devoid of chlorophyll in all its parts. The colour of the plant is usually yellow. In some cases plants take on an orange colour and some species have a reddish or purplish tinge. The flowers are very small and occur in dense clusters. The plants are annuals reproducing by seeds.

The seeds germinate on the ground in spring and the young seedling is able to lead an independent existence only as long as the stored food in the seed is avail-

Cuscuta on *Zizypus jujuba*

4

able. If the young *cuscuta* vine fails to come in contact with a congenial host plant about which it can twine, it soon perishes. In case it does find a host it soon establishes a parasitic relationship and continues growth until seeds are matured. The entwining vine puts forth parasitic roots in the form of suckers or haustoria which penetrate the tissues of the host and connect with both the water-carrying and food-carrying channels of the host stem. As soon as contact is thus established with the host the root and lower portion of the cuscuta vine perish, thus severing all connection with the soil and henceforth the parasite grows entirely upon the host plant and is wholly dependent upon it for sustenance.

Whenever and wherever found they should be collected before seeding and burnt. *Cuscuta* seeds will live in the soil for four or five years.

Mistletoe

Mistletoes are shrubs with elongate slender branches; the leaves are opposite, narrow, tomentose, olivaceous, with short petioles. The flowers occur at the axils of the leaves and are solitary, slender, cylindric, of brownish-yellow or yellowish-white colour. They form seeds in berries which are surrounded by a sticky gelatinous matrix which enables them to cling tenaciously to the bark of trees. Bird eat the berries and are probably responsible to a certain extent for dissemination. In some species the seeds are forcibly ejected from the berries at maturity and stick to the bark of branches wherever they happen to lodge. On germination the hypocotyle forms a disc-like appressorium which flattens out against the host. From the centre of this disc a root-like sinker penetrates the tissues of the twig and establishes connection with the water-carrying tissues. When established cortical roots radiate in the bark of the host and send additional sinker into the wood.

Mistletoe on a branch of Mango

The first noticeable effect upon the host after a mistletoe plant becomes established upon it is usually though not always swelling at the point of attack. Affected limbs first form fewer and smaller leaves;

fruits which may be formed are dropped and the limb becomes unproductive. With time the limb dies back from the tip and the parasite assumes the form and appearance of the branch. Large branches and even trunks may be greatly deformed and the younger trees entirely killed.

They are commonly found on mango, jack-fruit and other trees. The affected plants may be treated by breaking or scraping off the mistletoe plants and by pruning off small infested branches.

Broomrape

Broomrape (*Orobanchae*) is usually observed on tobacco, brinjal and mustard plants. It is a small flowering plant of whitish colour and devoid of green leaves. There are no roots, but the underground portion of the stem sends out root-like organs called *haustoria* by means of which the parasite attaches itself to the roots of its host plant. The stem is unbranched and thick and succulent at the base. The greater part of it bears white flowers tinged with blue.

Its seeds are very small, like dust particles and are produced in large quantities from a single plant. The parasite reproduces by seeds and grows only in the presence of its host plant. The seedlings on becoming united with the host plant draw nourishment from it through *haustoria*. The host plant being thus deprived of its food is left to starve and in cases of severe attack the affected plants are killed.

The seeds can remain dormant in the soil for a long time. The parasite should be prevented from maturing its seed. The plants should be pulled out as soon as they come above ground and should be burnt at once. Rotation should be practised and susceptible hosts not grown on the field for a few years.

Orobanchæ on Tobacco

Striga

Striga grows as a parasite on the roots of sugarcane and jowar plants. Its roots are closely connected with those of its host and by means of them it sucks up food from the latter. The affected plant loses much of its nutriment in this way and cannot grow properly. A large number of the parasite is generally found growing on a single jowar or sugarcane plant and each gives out 15 to 30 shoots. Badly attacked fields have a great many plants of the crop killed at an early stage and present the appearance of drought-stricken areas.

Striga has small narrow greenish leaves and grows to a height of 1 to 2 feet. It bears white tiny flowers which can be seen only after careful observation. If both the host and the parasite are dug out completely the connection of the roots of the two plants can be easily observed.

Striga first appears above the ground in July-August. Flowers come out in September and the parasite matures its seeds in small capsules in December. The seeds are very minute and get dispersed all over the field on the bursting of the capsules. Innumerable seeds of the parasite are in this manner dropped into the soil every year. Next year when the host crop is grown in the same field, the parasite comes up and attacks the crop.

In order to control the pest the parasite should not be allowed to form seeds. Before the formation of seeds they should be pulled out and burnt. Secondly, the field which has formerly borne an affected crop should be sown with a crop that is not susceptible to its attack.

(d) *Nematodes*

Amongst the animal parasites several such as insects, mites and microscopic animals (*nematodes*) live

Striga on Sugarcane

as parasites on plants. Insects and mites are serious plant pests. But these lie within the province of the entomologists. The nematodes are the cause of root troubles of scores of economic crop plants in nearly all parts of the world. Nematodes are so small that it is impossible to detect them without the use of the microscope. They lay eggs about 1/250 of an inch in length. Within the egg the young worm develops, eventually escaping by the rupture of the shell. It is then about 1/75 of an inch in length.

The nematodes have tubular or filiform bodies with mouth and well-developed alimentary canal. The mouth is provided with papillae or lips or with hooks or spines in the oral or buccal cavity and leads to a narrow oesophagus, which usually has thick muscular walls and a cuticularised lining and may be dilated into one or more muscular oesophageal bulbs or pharynx. By valves and muscular walls, the oesophagus pumps in fluid food as well as, in some species, solid particles and by peristaltic action, passes it on to the intestine, usually a straight tube which opens near the posterior end of the body on the ventral surface.

The body wall is muscular and encloses a body cavity containing the blood fluid, the alimentary canal and the excretory and reproductive organs. There is no definite circulatory system and respiratory organs are lacking. The body is unsegmented but the stiff cuticle is often ringed. The muscular body wall makes it possible for the body to be knotted, curved or bent and permits the characteristic undulatory movements.

The sexes are generally separate and the males smaller than the females. The females lay eggs but in a few cases they may bear living youngs. Many species are parasitic during either all or part of their life cycle, and some animal parasites require two entirely unrelated hosts to complete their life cycles.

The plant pathologist is interested in but one

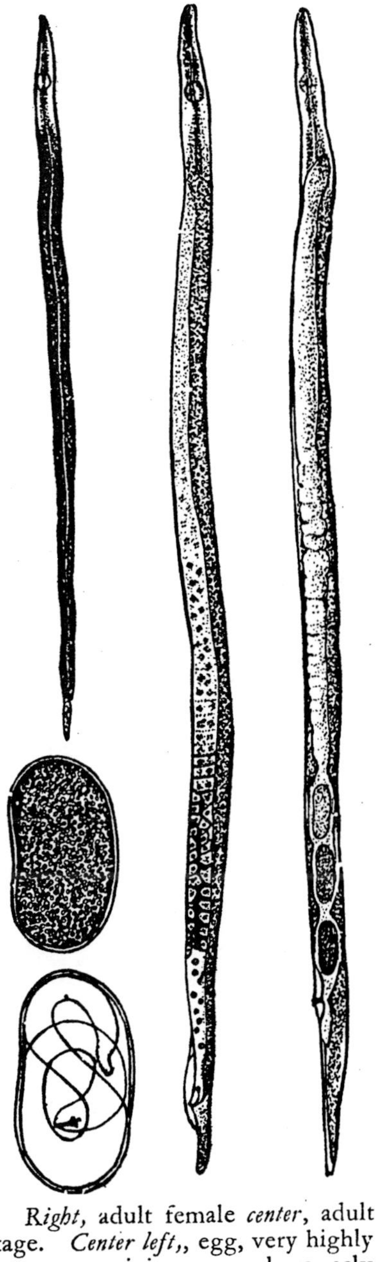

Nematode. *Right,* adult female *center*, adult male *upper left* larval stage. *Center left,*, egg, very highly magnified. *Lower left*, egg containing an embryo eclvorm, very highly magnified (After Owers)

family of the nematodes, the *Anguillulindae* (*Tylenchidiae*). This is a family of minute forms which live either free in the soil, decaying organic matter, water or on or in plants as parasites. The nematode disease is not serious upon all species of plants, but it is no doubt injurious to some extent. In several instances no external evidences of the presence of these worms in infested plants are visible. But on examination of the roots of the plants a number of galls may be found. If these galls are cut across and examined under the microscope myriads of these tiny worms will be discovered.

II. Non-Parasitic Diseases

(*a*) *Deficiencies of Soluble Salts*

Every green plant requires ten different chemical elements for its proper nutrition. These elements are: carbon, hydrogen, oxygen, nitrogen, phosphorus, sulphur, potassium, calcium, magnesium and iron.

Carbon, hydrogen and oxygen are combined to form carbohydrates, which are essential foods for crop plants. In addition to these three elements, nitrogen, sulphur and phosphorus enter into the composition of proteins and nucleoproteins which are manufactured by green plants. Potassium is essential for healthy growth and accompanies and plays a part in carbohydrate synthesis. Calcium is necessary for normal leaf development; it exists as calcium pectinate in the middle lamellae which cement adjoining cells and may serve a protective action by combining with oxalic acid to form crystals of calcium oxalate which are insoluble. Magnesium, if not an actual constituent, at least accompanies certain proteins and is contained in chlorophyll. Iron, in minute amounts, is essential for green plants and its lack prevents chlorophyll formation.

Elements Likely to be Deficient

Of all the essential elements cited above those which are most likely to be deficient in certain soils and to limit plant growth or give rise to abnormal or diseased conditions are nitrogen, phosphorus and potassium. Under certain conditions sulphur, magnesium, calcium and iron may be lacking in the proper amount. The absence of an element or its presence in non-available form will lead to diseased conditions.

(i) *Nitrogen shortage*—The usual symptoms of nitrogen shortage on all plants are: restricted growth of both tops and roots; shoots are short and thin, and growth is upright and spindly; leaves are small, usually of pale, yellowish green colour in the early stages of growth, and later they develop highly-coloured tints of yellow, orange and red, or sometimes purple; tinting begins on the older foliage and proceeds towards the younger leaves and the tints may also be developed in the petioles; defoliation is premature and begins at the older leaves; lateral shoots are few and lateral buds may die or remain dormant; blossoming in severe cases is greatly reduced and consequently yields of grain and fruits are very small.

(ii) *Phosphorus deficiency*—Phosphorus deficiency produces effects which are in many ways similar to those due to a shortage of nitrogen. Thus growth of both tops and roots are greatly restricted; shoots are short and thin, and growth is upright and spindly; leaves are small and defoliation, starting at the older leaves, is premature; lateral shoots are few in number and lateral buds may die or remain dormant; blossoming is greatly reduced with consequent poor yields of grain and fruits; the opening of buds, leaves and blossoms in the spring are delayed.

All the above points, it will be noted, apply equally to deficiencies of nitrogen and phosphorus. There are, however, points of difference. Thus with phosphorus deficiency, leaf colour is generally a dull, bluish green and the tints developed are usually purple rather than yellow or red, or they may take the form of a dull bronzing with purple or brown spotting. Leaf margins may also show brown scorching effects. Phosphorus deficient fruits are unlike those resulting from nitrogen shortage. They have a green ground colour and may be highly flushed; the flesh is soft and the fruit is acid in flavour and its keeping quality poor.

(iii) *Potassium deficiency*—Symptoms of potassium shortage may vary according to the extent of the shortage of the element. Thus in a mild deficiency in cereals, plants with rather thin shoots may be developed and, where the deficiency is acute shoots may die back and stunted plants with numerous tillers but few or no flowering stems, result. In trees, a slight deficiency leads to somewhat restricted shoot growth, and the shoots are thin, whilst when severe no terminal shoot growth occurs and shoots or entire branches die back, and eventually the tree may die. Leaf symptoms are striking and though somewhat varied are generally very characteristic for the different kinds of plants. The colour of the leaves may be dull bluish green, and there may be some chlorosis, especially in intervenal areas. One of the following symptoms will invariably occur, beginning on the older leaves; browning of the tips (tip burn), marginal scorching (leaf scorch), the development of brown spots, usually most numerous near the margins. The laminae of the leaves of many broad-leaved plants show a tendency either to curl backward toward the under surfaces or roll forward towards the upper surface parallel with the midrib.

Considerable stunting usually follows from potassium deficiency; internodes of stems are short, the production of grain and fruit is greatly curtailed and root systems may be very poor, trees often being loose in the soil.

(iv) *Magnisium deficiency*—Symptoms of this deficiency are most marked on the foliage, and the effects always developed initially on the oldest leaves and proceed systematically towards the younger ones. Chlorotic effects are common, which is not surprising in view of the fact that magnesium is a constituent of chlorophyll. The visible effect of magnesium shortage varies from a mild or faint chlorosis to a pronounced chlorosis and in the more severely affected plants may cause internal or marginal necrotic areas or complete necrosis of the leaves, the symptoms developing progressively from the base upwards. The chlorotic pattern follows the type of leaf venation showing (a) in monocots like maize and other cereals as pale or chlorotic streaks between the parallel veins; and (b) in many dicots as a rather characteristic mottle in which the chlorosis is confined to the intervascular areas, in the extreme types little green tissue remaining except the veins. Severely affected leaves may wither and be shed or may absciss without the withering stage.

(v) *Calcium deficiency*—The outstanding symptoms of calcium deficiency appear in the young leaves near the growing tips and in the root systems. The young leaves may be severely distorted and may either be rolled forward along the margins or curled backward towards the under surfaces. The leaf edges are irregular in form, often appearing ragged, and they may show brown scorching effects or thin, chlorotic marginal bands. Root systems are invariably poorly developed, remaining stumpy and failing to elongate, and they may appear gelatinous.

(vi) *Iron deficiency*—Iron deficiency is much more common in fruit plants than in vegetable or agricultural crops. The main symptom is always severe chlorosis of the leaves, and young growths are always most severely affected. In some instances the leaves may be almost wholly bleached and in others the chlorosis may produce a mottled pattern. Scorching of leaf margins and tips may occur in extreme conditions.

Therapeutic value of some rarer Elements

In addition to the ten elements generally recognised as essential in the nutrition of plants, certain others which occur in the soil in relatively small amounts may have a corrective or therapeutic value for certain pathological conditions. The exact way in which this corrective effect is accomplished is in many cases uncertain. Some of these trace or rarer elements such as boron, copper, manganese, zinc and others have been used effectively for the correction of certain diseases of our crop plants. Some of the trace elements occur in plant tissues so constantly that a physiological significance is suggested.

Boron has been successfully employed in curing 'bog disease', a trouble characteristic of swampy heath soils in European countries. Some of the crops affected are barley, oats, wheat, clover and turnips. Barley shows yellowing and marginal curling or rolling of leaves and empty heads; oats, a whitish colour with increased stooling; wheats, longitudinal white stripes on the leaves and heads rudimentary or lacking; clover, leaf pallor, dry brown spotting and marginal chlorosis, with flowers late or scanty; turnips show stunting, discolouration and shriveling of leaves, with yellowish white lesions spreading in from the margin. The curative value of boron has also been demonstrated for the brown heart disease of turnips and swedes, cork and drought spot of apples and the crack stem of celery.

In certain cases of chlorosis copper has been shown to have a beneficial effect whether applied in the soil or sprayed on the leaves. Growth abnormalities of many plants that are produced on peat and muck soils have been corrected by the application of copper compounds. Copper additions produce better colour of such crops as onions, spinach, lettuce and carrots, increase sugar content of beets and carrots and improve the flavour of most crops. The addition of copper to acid mucks prevents the premature dying of onions.

The upper leaves of tobacco plants, grown without copper, are unable to maintain their vigour and wilt badly. The leaves wilt prematurely and do not gain turgidity even in the presence of sufficient moisture. It has also been observed that the amount of seed set is reduced and the seed stock is unable to stand erect if copper is limiting.

In the absence of sufficient manganese, tomato, bean, oat and tobacco are dwarfed. Associated with this dwarfing is a chlorosis of the upper leaves of the plant, and the leaves become spotted. The tissues of the spots die and frequently drop out. The 'gray speck' of oats has been attributed to a shortage of this element in some soils. Some of these troubles at least are prevented by spraying or dusting with manganese sulphate or by soil treatment with manganese sulphate.

Pecan rosette, the yellows of walnut trees, the mottle leaf of citrus, the little leaf of stone fruits and grapes, white bud of corn and the bronzing of the leaves of tung trees are all ascribed to zinc deficiencies. In tobacco plants, a zinc deficiency is characterised by a spotting of the lower leaves and in extreme cases almost total collapse of the leaf tissue may occur.

(b) *Excess of Soluble Salts*

Certain types of growth are directly traceable to overnourishment, the first effect of which is to be seen in an increased vegetative development, a deeper green than normal, more succulent tissues and a retardation or suppression of reproductive functions. Generally overnutrition leading to an excessive accumulation of plastic substances in the plant in proportion to their utilization may lead to pronounced morphological changes. Among these changes may be mentioned *phyllody*, or the transformation of floral organs into leaf-like structures; *petalody*, or the transformation of calyx bracts into petals; *pistillody* or the change of stamens into carpels.

(i) *Excess of nitrogen*—Some of the injurious effects of too much nitrogen are: delayed maturity, dropping of flower buds, a succulent growth with poor mechanical resistance, chlorosis of foliage, followed by burning or necrosis, browning or corrosion of the roots, gummosis and die back in citrus or stone fruits, lowering of quality in fruits or cereals and increase of susceptibility to parasitic diseases.

(ii) *Excess of lime or manganese*—In many cases the calcareous or manganiferous soil cause chlorosis by iron starvation. Sufficient iron is present but in lime chlorosis the excess of lime changes the iron compounds into insoluble colloidal iron which cannot be utilized by the plant, or in the manganiferous soils to a much more difficulty available ferric iron.

(iii) *Excess of boron*—Boron in small quantities may be beneficial but excess causes retardation or prevention of germination, death of plants or stunting so as to give imperfect and uneven stands, absence of normal colour marked by the bleaching or yellowing of normally green parts, especially the margin and tips of leaves, reduced growth and premature ripening with lessened yields.

(iv) *Excess of acid*—The first effect with moderate acidity will be retarded growth and a pallor or less intense green than normal, which if the unfavourable conditions continue may become more pronounced. Foliage may become mottled, showing lighter green areas between the veins, or the chlorosis may become diffuse or general. Such affected plants may weaken and die prematurely or growth may be resumed if rain modify the acidity later in the season. Roots make a poor development and many of the lateral feeders may be repeatedly killed back.

(v) *Excess of alkali*—The effects of alkali will vary according to the concentration and kind of the salts present and the resistance or tolerance of the plant to alkali salts. Some of the effects are: failure of seeds to germinate, death of seedlings after reaching a few inches in height; retarded germination, the formation of sickly slender plants marked by chlorosis and early death without flowering or fruiting; retarded growth, chlorosis and some foliage burning but fruit ultimately reaching maturity. Established shade or fruit trees may show a retarded growth or dwarfing, short terminal shoots, leaves fewer and smaller than normal, and frequently showing much chlorosis, burning or blighting of leaves at the tip or margins and premature leaf fall. Black alkali may cause a corrosion of the bark at the crown.

(c) *Deficiencies or Excess of Water*

The importance of a proper amount of soil moisture in crop production cannot be overestimated. Of the various factors affecting crop production the most influential has been recognised to be water supply. Water is one of the external factors in plant growth in that it is necessary in the processes of weathering. It also functions as an internal factor in plant development, inasmuch as it maintains the turgidity of the plant cells, acts as a carrier of food materials, functions as a regulator, and can actually be utilized as a source of hydrogen and oxygen. But its deficiency or excess in the soil is detrimental to the proper and healthy growth of plants.

(i) *Effects of moisture deficiency*—Shortage of moisture is marked by yellowing, reddening or other discolourations followed by leaf fall in woody plants. In plants suffering from drought, dead brown areas may appear in the intercostal areas of leaves, in the centre of areolae of these areas, or the leaves may be blighted or burned at the margins or tips. Moisture shortage by its interference with nutritive processes may lessen the production and storage of reserve food. Tuber and root crops will remain small, and cereals will produce shriveled grains. Fruits may be spotted, deformed or under normal in size or they may shrivel and fall prematurely.

(ii) *Effects of excess Moisture*—Excess of moisture is marked by yellowing and decomposition followed by decrease in production. An abundant water supply produces a type of growth that is more susceptible to the inroads of either bacterial or fungus pathogenes. The greater delicacy of the cell walls in plant structures provided with an abundance of water and the increased rate of growth frequently result in the rupture of organs, such as fleshy roots, tubers, stems and fruits. The fall of leaves, the shedding of blossoms, the dropping of fruits or the casting of twigs may sometimes result from a disturbed water relation, either an excessive supply, a shortage or abrupt fluctuation.

(d) *Improper Air Relations*

Living plants stand in intimate relation to their air environment. Unfavourable air relations may interfere with their normal physiological processes and growth and cause disease. *Blackheart* of potatoes, *black leaf speck* of cabbage, cauliflower and other crucifers, *brown heart* of pears and apples and a number of other diseases have been ascribed to improper air relations. The general effects of the different factors which make up the air environment on plant growth are listed below.

(i) *High Temperatures*—The principal types of heat injury are: retarded growth and undersize or failure to mature the flowers and fruit; localised killing of tissues or a sunburn or sunscald of leaves, flowers or fruits; localised killing of stem tissues or the formation of heat cankers; defoliation or premature falling of leaves; premature ripening of fruits and death of the plant as the result of a heat necrosis.

The scalding of flowers frequently results from high temperatures. Many fruits suffer from sunscald or sunburn, especially those which are succulent. Even fruits like the apple may be severely burned during periods of intensive summer heat.

Heat defoliation may occur in both deciduous and evergreen trees. Seedling of deciduous and herbaceous species may be seriously injured or even killed by high temperature, the injury being localised in the stem just above the ground level, thus resembling damping-off due to fungi.

(ii) *Low Temperatures*—For each species, variety or strain of plant there is a *minimum* temperature at which growth first begins, an *optimum* temperature at which the growth is most rapid and a *maximum* temperature beyond which growth ceases. As the temperature sinks below the optimum, growth becomes less and less rapid and finally ceases at the minimum. This retardation or checking of growth is the inevitable result of low temperature. A second effect of low temperature is the prevention of chlorophyll formation or the slower formation of this pigment, with the result that parts normally green may become yellow. In some plants or plant parts, cold causes the development of red pigment, which apparently obscures the lesser degree of chlorophyll development. When the temperature sinks to a sufficient degree freezing of plant tissue results and death may follow or with the rise of temperature the frozen tissues may thaw out without any appreciable injury.

(iii) *Frost injury*—The effects vary with the degree of cold and the sensitiveness of the plant and include: chlorotic bands or spots or general chlorosis or yellowing; the formation of red pigment, *anthocyanin,* in organs normally green; blistering, crinkling, curling puckering, irregular laceration or shot-holing of leaf blades, sometimes with reduction in size; and the complete necrosis or blighting of leaves and vegetative shoots.

Blossom buds or blossoms may blight and drop without setting fruit or late frosts may affect fruit after it has already set and cause shedding, reduction in size, malformation, internal necrosis or sterility. The essential organs of the flower, stamens and pistil are generally more sensitive than the accessory parts.

(iv) *Light deficiency*—The *optimum light income* denotes the amount of light which will induce the best growth or produce a type of development that is most nearly normal. As the light income drops from the minimum to zero or complete darkness, the plant may undergo gradual formative or structural changes, including alteration of colour and peculiarities of structure which are characteristic of the condition known as etiolation.

The changes which are characteristic of etiolation are abnormal elongation of stems (internodes) and petioles or of leaves; pronounced reduction in the size of the leaves; reduction in the amount of chlorophyll or its complete disappearance (in darkness), suppression of the reproductive function; a soft or succulent type of growth.

(v) *Effect of intense light*—With increase in the intensity of light photosynthesis will increase up to a certain point and then with further increase in intensity of light, the photosynthetic activities remain about constant, but for a short time only. If the intensity becomes too great or the optimum is exceeded for too long a period, the construction of carbohydrate food becomes less active and may finally be checked entirely. Under conditions of intense illumination, sensitive plants develop a pale or yellowish green cast or even a bronzing of the leaves.

The sensitiveness to intense light often varies with the age of the plant, for example the seedlings of many trees are not able to withstand the direct sunlight, while older trees suffer no derangements from exposure to the same light intensity. In the more sensitive species continued exposure to intense light may kill the protoplasm in leaves, stems or fruits and browning, burning or blighting of localised spots or more extended areas may be the final result.

III. Virus Diseases

Virus diseases of plants constitute one of the great groups of plant maladies which are entirely distinct and different from those caused by visible parasites or non-parasitic influences. These troubles have, however, one feature in common, namely, they may be communicated from diseased to healthy plants by an infectious principle, the *virus*.

Virus diseases of plants have during recent years attracted much attention. This is scarcely surprising for large losses on many crops are sustained annually as a result of their attack. In 1886 Mayer described and proved the infectious nature of a disease of tobacco which he designated as '*Mosaickrankheit*'. Six years later Iwanowski showed that the causal agent would pass through the procelain bacterial filters. This was a new and startling discovery. Previous to this time such disorders had no doubt been attributed to physiological disorders or some environmental disturbance.

Symptoms

The foliage shows the symptoms of the disease first and the young foliage will usually show it much more pronounced than the older parts. In general the different types of symptoms attributable to viruses may be classified as follows:

(i) *Chlorosis*—Most of the virus diseases cause a change in the chlorophyll content of the leaf which is manifest in the different intensities of colour. The chlorosis may be general or a localised mottle or mosaic of the foliage. A general paling of leaves or other green parts with the final production of a rather *general chlorosis* or *yellowing* is a very pronounced symptom in certain virus diseases. On the other hand the mottling of the foliage due to alternating patches or spots of light green or yellow and dark green is characteristic of a very large number of virus diseases known as *mosaics*. The mosaic mottle may be very mild or only evident as slightly paler green spots or it may be very pronounced with the extreme represented by distinctly yellow or even white spots. The spots vary in form and size from irregular to angular and from small to large, certain patterns being characteristic of specific mosaics.

(ii) *Distortion*—Distortion is one of the very common symptoms of many of the virus troubles. It consists of puckering, cupping, curling or rolling of the leaves in various ways.

(iii) *Necrosis*—Necrosis or browning and death of tissues is a marked effect of certain viruses and may involve localised groups of cells, either superficial or internal. The resulting external picture may be necrotic spots, streaks or extended necrotic areas finally involving entire leaves or portions of stems. Internal necrosis may involve the vascular tissue, especially the sieve tubes and companion cells.

(iv) *Dwarfing*—Reduction in size of leaves or other organs or of the entire plant is a feature of many of the virus diseases. Stunting with more or less rosetting is characteristic of bunchy top of bananas, stunt of rice, rosette of peanuts and others.

Types of Virus Diseases

There are, generally speaking, three types of virus diseases:

(i) *Mosaic diseases*—Symptoms are the mottling of leaves, with or without malformation. They are transmitted by mechanical and biological methods. Common examples are tobacco mosaic, potato mosaics, cucumber mosaic, legume mosaics and bean mosaic.

(ii) *Yellow diseases*—Symptoms are general chlorosis with clustering or upstanding habit of the young leaves. Transmissible by biological means only. Common examples are aster yellows, banana bunchy top, peach yellows and potato leaf-roll.

(iii) *Miscellaneous*—Various symptoms neither mosaic nor yellows. Transmission by either mechanical or biological methods or both. Common examples are little peach, witches' broom of potato, wheat rosette and spindle tuber of potato.

Theories about Cause

The problem of determining the exact cause of the virus diseases has been a baffling one. Many theories have been advanced to account for the disease, to mention, the unbalanced nutrition theory, enzyme theory, bacterial theory, protozoan theory and virus theory. The virus theory is most generally accepted though the exact state of existence of the virus is not known. The term '*virus*' as used today includes the existence of an ultra-microscopic organism or an infective principle of an unknown type in the expressed juices of an infected plant.

The term '*virus*' is borrowed from human and animal pathology. It has been used by animal pathologists for a long time to designate a contagium that cannot be definitely associated with a known organism. This infective principle seems to be too small to be seen by a microscope and in general not to be culturable by any of the bacteriological methods known.

When the infectious nature of the plant diseases of this type became known through the discovery that some kind of an infective principle is carried in the juices of diseased plants the first impulse of course was to look for bacteria as the causal agent. But when persistent effort failed to find bacteria the natural tendency was to note the striking resemblance between this type of plant disease and the virus diseases in animals and to adopt the term '*virus disease of plants*' for this group of plant diseases.

Properties of the Virus

While doubt still exists as to the exact identity of the causal agency in the virus diseases of plants, yet a great deal is known concerning various characteristics of this infective principle. It has been definitely established that the contagium is in the sap of the diseased plant and can be extracted by grinding up the diseased plant parts and expressing the juice. The extract thus secured can be used as an inoculum and the disease produced in healthy plants.

When a small quantity of the infective principle is introduced into a healthy plant, the virus is able to multiply or increase in amount. In this respect it resembles living organisms. It is readily transferable from the point of inoculation through the sap channels to other parts of the plant and as a rule after an incubation period of several days to a few weeks the virus is found in all parts of the plant.

The effects of many different chemicals have been

tried. In general it requires rather strong solutions of the various chemicals to kill the virus. Nitric and hydrochloric acids do not render the virus inactive in concentrations of less than one gram in 50 to 100 cc. It requires a concentration stronger than one gram of potassium permenganate or zinc chloride in 100 cc. of water to kill the virus. Alcohol stronger than 55 per cent. is highly toxic. Eighty per cent. alcohol kills the virus in less than one-half hour. Four per cent. formaldehyde is very toxic to the infective principle.

Temperatures near the boiling point quickly kill the virus. The virus is not so sensitive to low temperatures, being able to withstand a temperature of —180°C. without losing its infective properties.

A virus extract will often remain active for a considerable period after separation from the plant Ordinary tobacco mosaic virus will retain its activity for several years stored in a bottle but many viruses will not tolerate even a few minutes separation from their hosts.

Transmission

Virus diseases may be transmitted in different ways in nature or artificially as follows:

(i) *Seed*—Although most virus diseases are not transmitted by seeds from a diseased plant a considerable number find their way into the seeds which then produce seedlings, of which a certain percentage may be diseased. Bean mosaic is most readily transmitted by seed, but rarely reproduces more than 50 per cent. of diseased seedlings. Leaf-roll of potato, lettuce mosaic, cucumber mosaic, tomato mosaic etc., are transmitted by seed.

(ii) *Vegetative propagation*—All the cuttings, tubers, scions and divisions made from a diseased plant carry the virus. Vegetative propagation of plants provide conditions for the very rapid and complete spread of virus diseases.

(iii) *Contact*—The extreme sensitiveness is illustrated by tobacco mosaic which may be transmitted by contact or by touching a diseased plant and then touching a healthy one. It has been proved that workmen who chew tobacco can introduce the mosaic by wiping the mouth and then weeding and handling the plants or even by 'spitting in the bed'.

(iv) *Pollen*—Bean mosaic has been reported to be transmissible by means of the pollen. Flowers of the healthy plant were emusculated and fertilized with pollen from a diseased plant. Seeds produced from such a union gave seedlings of which a considerable percentage were diseased. Mosaic of Jimson weed is also transmitted by pollen.

(v) *Insects*—The activity of insects in spreading virus diseases is now well established and it is known that aphids, jassids, capsids, leaf-hoppers, grass-hoppers and other families are capable of playing this part. A particular species of insect may be capable of transmitting more than one virus disease. The aphis *Myzus persicae* has been shown to be capable of spreading no less than 14 distinct viruses. Some diseases, however, as aster yellows and banana bunchy top appear to be transmitted by only one species of insect.

(vi) *Soil*—A good deal of doubt still exists as to whether viruses can be transmitted through the soil although positive statements to this effect have appeared from time to time in the literature. On the whole, soil transmission probably plays a very small part in the general spread of plant viruses.

(vii) *Inoculation*—This means the actual application of the virus-containing sap from diseased plants to healthy plants. This method is successful in most of the mosaics and in some other virus diseases.

(viii) *Budding* or *Grafting*—There are some virus

diseases which cannot be transmitted by any other way than by budding or grafting e.g.. spike disease of sandal, potato witches'-broom, hop mosaic, all peach viruses except yellows.

Hosts

Since the early work of Mayer on tobacco mosaic, published in 1886, many different types of virus diseases have been identified and the list of host species known to be susceptible to one or more of them is constantly increasing in number. Members of the *Solanaceae* are especially susceptible and a number of them were among the first on which these diseases were studied. No attempt to give a complete list of susceptible species here will be made but a partial list is included to show the widespread prevalence of this kind of plant diseases. Among the families known to contain susceptible genera and species are the following: Solanaceae, Rosaceae, Cucurbitaceae, Leguminoseae, Umbelliferae, Asclepiadaceae, Ranunculaceae, Labiatae, Malvaceae, Euphorbiaceae, Chenopodiaceae, Compositae, and Gramineae. Some of the plants which are susceptible to one or more virus diseases are tobacco, potato, tomato, cucumber, pumpkin, raspberry, blackberry, strawberry, peach, sugar-cane, corn, spinach, celery, bean, cabbage, turnip, lettuce, clover and papaya.

Economic Effects

The economic significance of virus diseases varies greatly. They may cause from very minor to very severe damage to our economic crops. Tobacco may be depreciated about 55 per cent. in yield and quality. In sugarcane the losses have been reported as high as 70 per cent. in India. Sugar beet losses in America amount to 75 per cent. It is estimated that 1½ bushels of potatoes are lost for every 1 per cent. infection in fields.

CHAPTER IV

CONTROL AND PREVENTION OF PLANT DISEASES

The practical agriculturist is primarily interested in protecting his crops from diseases and usually cares little about the cause of the trouble provided he can find an efficacious remedy. The latter statement is true until he is made to realise that the most efficient control measures usually are based on accurate knowledge of the nature of the trouble. The more a cultivator knows of the causes and action of the disease the more intelligently and the more successfully will he be able to combat it. In olden days the remedies recommended were necessarily empirical since nothing was known of the cause of disease in most cases. Even today we are using many control measures that originated empirically however much we have modified and adapted them in recent years in accordance with our increased knowledge of the cause of the disease. With the gradual acquisition of more accurate information on the nature and cause of diseases we have worked out methods and devices to meet the special need in combating any particular disease.

Economic Importance of Plant Diseases

The number of plant diseases is very great, and even if those occurring on non-economic plants are ignored, there is still a large residue. A glance at the disease lists of economic crops such as are published from time to time by departments of Agriculture and other bodies will show as many as fifteen or so diseases occurring on particular species of cultivated plants. The majority of these appear for the present at any rate to be of little economic importance but one may say generally that more or less all cultivated plants are subject to two or three diseases of major importance which at various places interfere seriously with their profitable cultivation. Thus in the case of the potato one may cite four important diseases, namely, blight, scab, rot and virus diseases all of which offer serious problems to the grower. In any one year and in a given locality certain of these diseases may do little damage or may even appear to be absent but over a wide area and over a period of years the aggregate loss is very great.

Estimates have been given of the total loss due to plant diseases in various countries. In 1891 the loss of wheat, rye, oats and barley from rust in Prussia has been estimated at over three hundred million rupees. In Australia in 1890-91 the loss by wheat rust was estimated at 36 millions of rupees. Fungus disease has practically exterminated the coffee plantations of Ceylon where the loss from 1870 to 1886 was about 15 million rupees annually. The minimum loss India suffers annually due to wheat rusts has been calculated to be Rs. 40,00,000. The potato blight disease caused famine in Ireland in 1845-46.

Estimates made by the United States Department of Agriculture indicate that approximately one bean in every dozen, one apple in every seven, one peach in every eight, one bushel of irish potato in every twelve, and one bushel of wheat in every ten are destroyed annually by diseases in crops.

From what has been cited above it will appear that monetary losses due to plant diseases are of enough significance to justify an intensive study of their causes and the means of controlling them.

How plant diseases cause injury and losses

Types of injuries and losses caused by plant diseases are varied; the principal ones are briefly stated below:

(i) *Killing of plants*—Plants, both annuals and perennials, may be killed. In our annual crops such as cereals and other field and forage crops, garden vegetables, etc., killing during the juvenile stage may result in thin stands or even such a complete kill as to require reseeding or replanting. In other cases premature death may come at varying time before maturity but frequently sufficiently early to prevent harvesting of the desired crop. This may be illustrated by the killing of nearly mature *arhar* plants by wilt. The loss of perennial plants like orchard crops which require a long period to reach the producing stage is more serious. Fungus root rot may destroy an orchard in 3—4 years.

(ii) *Reduction in yield*—The effect of disease upon the yield of the marketable product may vary all the way from only a slight reduction to a complete failure without causing the death of the affected plants. Many cases of reduced yields as a result of disease pass unnoticed by the grower or the amount of damage is often underestimated. Cereals may suffer serious reduction in yields from smuts and rusts as well as marked reductions of grade or quality. Virus diseases cause appreciable reduction in the yield of potato.

(iii) *Reduction in productiveness of perennials*—Certain diseases of perennial plants may not be sufficiently virulent to kill the affected plants out-right but rather leaves them to grow in a dwarfed or crippeled condition for a period of years, sometimes completely ruined from the standpoint of production or in other cases with a greatly lowered yield, perhaps of inferior quality. Orange trees affected with chlorosis and die-back have been found practically worthless. Plum trees may be so severely infested with black knot that they are almost useless and in certain cases this disease has led to the complete abandonment of the culture of this crop.

(iv) *Lowering of grade or quality*—The marketable product from various crops may be lowered in grade or quality from various defects caused by parasitic organisms acting directly or indirectly or by injuries resulting from non-parasitic influences. Fungus lesions including scab, fruit spots, blotch, numerous types of rots or decay and rust may lower the quality of the plant product. The important non-parasitic defects which may appreciably impair the quality of the marketable product are: bitter pit, sunscald, frost injury, spray burn, hail injury, wind falls, bruises, cracks, under-colour, under-size, etc.

(v) *Destruction of the market product previous to harvest*—Fruits, vegetables and other crops represented by fruits, tubers, bulbs, corms and fleshy roots and other storage organs which are reserves of food material are subject to rapid decay. The decay, in a large number of cases, starts in the field before harvest during the maturing period of the drop and may continue in the harvested crop and cause heavy losses during storage and during transit to market. Some of the rots of potato, brinjal, gourd, and apple begin in the field previous to harvest but become serious during transit and in storage. The smut diseases of cereals, especially of wheat, oats and barley offer striking illustrations of the destruction of the market product, the grain, previous to harvest.

(vi) *Losses in Storage or in transit*—Vegetables and fruits usually suffer serious losses during storage and in transit. Soft fruits and the succulent or fleshy vegetables suffer the heaviest losses. In potatoes storage and transportation losses may be heavy due to blight (*Phytophthora infestans*) and *Fusarium* rots;

in sweet potatoes due to soft rot; in cabbage, peppers, carrots, beans, lettuce, etc., due to bacterial soft rots (*Bacillus carotovorus*); in citrus, pineapple, bananas and other sub-tropical fruits due to species of *Penicillium*, *Ceratostomella*, *Gloeosporium*, etc.

I. Control of Parasitic Plant Diseases

For the successful control of parasitic plant diseases a knowledge of the life history of parasitic organisms is of fundamental importance. A knowledge of the life history of parasites makes it possible to discover weak or vulnerable points, if any exist, and to strike at the proper time and place. Without a knowledge of life histories we might still be trying to control smuts by spraying, rusts by seed disinfection and downy mildews by soil disinfection. It is true that some very important advances have been made in disease control by accident but during the more modern development of plant disease studies, effective control has depended more and more on planned scientific investigations.

Some of the methods adopted for controlling parasitic plant diseases are listed below:

(i) *Sanitation*—A great many plant disease organisms winter over or are perpetuated for a short time a least on various forms of plant debris. The brown rot and bitter rot fungi hibernate on rotted and mummified fruits. Many fungi and bacteria survive in the prunings that are left in the soil. All such diseased debris should be burnt or destroyed in some way.

(ii) *Crop rotation*—Many pathogenic organisms harboured in the soil are parasites to certain species of crop plants. If these certain crops are not grown in a field for a few years the special parasite of the crop will die out in that field. It will then be safe to grow the susceptible crop again in this soil. The length of time an organism will survive in the soil in the absence of its host plant determines the length of the rotation cycle.

(iii) *Surgery*—(Cutting out diseased parts). Many of the canker diseases, e.g., fire blight, European canker and apple tree anthracnose can be fought more or less successfully by cutting out the diseased areas. Such parts should be destroyed by burning or by other means after separation from the host.

(iv) *Roguing*—(Uprooting and removal of diseased plants). This method is usually practised with truck and vegetable crops. It can be successfully used against those diseases which can be easily recognised in the field during the growing season, for example, wilt diseases of potato. Such removal completely or partially checks the further progress of the disease.

(v) *Removal of alternate host*—Fungi which are heteroecious complete their life cycle on two different hosts. Wheat rust for example is known to pass a part of its life on wheat and the rest on barberry in some countries. In such cases one means of controlling the disease consists in the eradication of the alternate host.

(vi) *Destruction of weed hosts*—Many parasites causing disease of economic plants also attack other related plants which are weeds or of no economic importance. The *Rhizoctonia* fungus which attacks a large number of plants is also known to attack a wide range of weeds. The destruction of these weeds or non-economic hosts is to be recommended as one means for combating these diseases.

(vii) *Sterilization of soil*—Plant disease organisms which live in the soil can be controlled by sterilizing the soil. Soil sterilization can be done by steam, heat or chemicals. The method is intended to kill the living parasite in the soil.

(viii) *Sterilization of tools*—In cases where diseased parts of plants are cut off it is recommended that the tools be sterilized between cuts to prevent carrying the germs from diseased tissues and depositing them on healthy tissues or plants. During pruning of trees many healthy plants get diseased, the disease-causing germ being taken from diseased plants to healthy ones through the pruning instruments.

(ix) *Disinfection of containers*—In cases where the pathogene in the form of spores or bacteria separates from the host and adheres to containers, carriers or storage bags or bins, there is danger of contaminating the next batch of products placed in such containers. Therefore the containers or carriers should be disinfected before filling again. This will apply to grain sacks which have held smutty grains, boxes or crates in which produce has moulded or decayed and railway cars.

(x) *Disinfection of wounds*—Wounds that occur naturally or due to pruning or removal of diseased parts in plants should be covered with a fungicide in order to avoid infection by fungi or bacteria. Many fungi and bacteria make their entrance into plant body through wounds.

(xi) *Seed disinfection*—Disinfection of seed and other propagating parts of plants are important methods in controlling plant diseases. Seeds may be treated by hot water or fungicides before sowing; this will kill the living spores or mycelium of the parasite in the seed without injuring the viability of the seed.

(xii) *Changing the soil reaction*—The acidity or alkalinity of the soil sometimes exerts a strong influence upon plant pathogenes which live in the soil or attack the under-ground parts of plants. If a certain organism thrive best in an acid soil it may be combated by changing the reaction of the soil by the addition of lime. If it is an organism which is favourably influenced by an alkaline soil the control procedure may be reversed and the soil treated to render it less alkaline.

(xiii) *Application of fungicides*—Fungicides are used to a considerable extent for controlling plant diseases. Fungicides are of greatest benefit in combating parasites which live on the surface of the host plant i.e., the *ectoparasites*. But the plants can also be protected from the attack of air-borne parasites by the use of fungicides. The object aimed at here is to cover the surface of the plant with a thin but more or less continuous film of the fungicidal substances so that when the air borne spores of the parasite arrive they are unable to initiate attack.

(xiv) *Growing of resistant varieties*—Some varieties of plants are less susceptible to a specific disease than other varieties. World over plant breeders are constantly trying to find disease resistant varieties by breeding and selection, Some varieties of wheat are known which are more resistant to rusts and smuts than other varieties. Other things being equal the more resistant varieties should be grown.

(xv) *Selection and growing of clean seed*—Where disease on seeds can be detected readily enough it is sometimes possible to sort out the diseased seeds and separate them from the healthy ones. Bean seeds affected with anthracnose or potato seed tubers affected with scab may be easily sorted out. But this method is not entirely satisfactory as some of the seeds which are slightly affected are likely to escape notice. But beans or other seeds may be safely collected from a field or locality where the disease does not occur.

II. Control of Non-Parasitic Diseases

Non-parasitic diseases of plants which are due to unfavourable physical and chemical conditions of the soil or due to unfavourable atmospheric conditions can be mitigated by modifying the environmental factors.

The soil characters can be changed to a certain extent but it is almost impossible to change atmospheric condition.

Unhealthy water relations can be controlled by ploughing, addition of organic matter, green manuring, irrigation and proper drainage.

Diseases which are due to lack of certain nutrient elements can be corrected by supplying these elements in available forms in the shape of fertilizers and salts. Troubles that are due to excess of certain elements or salts can be corrected by applying substances that will neutralise their effects.

III. Control of Virus Diseases

The control of virus diseases is even more difficult than the parasitic and non-parasitic diseases. Since a large number of the virus diseases are transmitted by insect vectors, theoretically, control should be possible by their elimination or extermination, but this is rarely possible. As these troubles usually spread from a few infected plants early removal and destruction of infected plants or trees may be helpful in protecting the balance of the crop from infection. Removal of weeds and their destruction may control or keep under control a number of the virus diseases which are known to be found on weeds. Use of disease-free seeds and propagating parts is another sound way of avoiding these troubles. Disease resistant plants appear to give most promise of control and in the case of sugarcane, bean, spinach and cucumber resistant strains have been developed but in others no resistance has so far been discovered.

In brief some of the important methods to be followed in controlling virus diseases are: prevention of mechanical transmission, eradication of weed hosts, use of disease-free seeds and propagating parts, roguing out of diseased plants, rotation of crops, indexing of seeds, spraying for the control of insects and growing of resistant varieties of plants.

CHAPTER V

FUNGICIDES: SPRAYING, DUSTING AND PAINTS

The term *fungicide* broadly considered is any substance used to kill or to prevent the growth of fungi including bacteria. According to their mode of action, it is possible to classify fungicides into two groups: *direct* and *protective*. The direct fungicides kill through contact with the fungus on the plant surface while the protective fungicide prevents the establishment of fungal infection.

Fungicides, to be effective, must fulfil two conditions. Firstly, it must be effective against the fungus against which it is applied, yet should cause no injury of economic importance to the host plant. Secondly it is necessary that the toxic material be amenable to application in an effective manner; the direct fungicide must make contact with the fungus to be destroyed, the protective material must leave a deposit of satisfactory protective properties. The methods of application of fungicides to the growing plant are conveniently classed on a physical basis thus:

(i) *Spraying*—When the toxic agent is applied in liquid form as a solution, suspension or emulsion.

(ii) *Dusting*—When the material is applied in the form of a finely divided powder.

I. SPRAYING

The grower should bear in mind that spraying is not a panacea for all troubles but only one of the important measures for use against certain plant diseases.

Spraying in the sense here employed consists in the application of fungicide in water as a carrier in the form of a fine mist to the aerial parts of plants. Any chemical compound or mixture selected for the prevention or treatment of plant troubles by spraying should fulfil the following specifications:

(i) It should be sufficiently toxic or strong to have a killing or inhibiting effect upon either dormant spores, germinating spores or upon mycelium.

(ii) The ideal spray should not injure or disfigure the plant surfaces to which it is applied.

(iii) The spray should stick to the parts to be protected and should cover the surfaces with a uniform film which should not be easily washed off after drying.

(iv) The active principle of the fungicide should be soluble or capable of change to a soluble or diffusible form under the influence of atmospheric conditions.

Materials for spraying

Very many substances have been used for spraying in the past and the number still employed in certain localities and under specific conditions for certain valuable crops, such as the potato, is considerable. For practical purposes the list may be reduced to a few that have proved applicable under very varying conditions and that have established themselves as efficient when properly made and employed, in a large number of cases.

(i) *Bordeaux mixture*—This is the earliest, the most widely known and the most generally useful of all spray fluids employed against parasitic fungi. The ingredients of bordeaux mixture are: copper sulphate, lime and water. The formulae used for preparing the mixture vary considerably and it cannot be said that there is any general agreement as to the best to use. The chief strengths recommended are:

(a)	Copper sulphate	4 lbs.
	Quick lime	4 lbs.
	Water	50 gallons

This may be selected for general use on fruit trees (except peaches) and on most field, garden and planters' crops.

(b)	Copper sulphate	5 lbs.
	Quick lime	5 lbs.
	Water	50 gallons

This is much used in general spraying and is often known as one per cent. Bordeaux. It is commonly used in spraying potatoes against blight. Weaker strengths of bordeaux, for example, 2:4:50 or 3:6:50, may be used for spraying more delicate plants.

Preparation of Bordeaux mixture—In making bordeaux mixture solutions of copper sulphate and of the lime should be made up separately. Dissolve the copper sulphate in half the water, in a wooden vessel by suspending it in a piece of gunny sacking just immersed. It dissolves slowly and may conveniently be put in overnight. Then slake the lime in a second vessel with a few pints of water added little by little until the bubbling ceases, after which pour in the rest of the water. Then mix the two solutions together either by pouring one into the other or the two together into a third vessel. Pour the lime through a strainer to keep back all lumps and stir the mixture all the time while pouring. Stir and strain again when pouring into the sprayer.

Bordeaux mixture as generally prepared is alkaline, containing an excess of lime and turning red litmus paper blue. An excess of copper compounds is dangerous to the foliage of many plants and is indicated by the solution turning blue litmus paper red. Another way of determining if the mixture contains an excess of copper is to immerse a bright iron or steel surface, such as a knife blade or a nail, in the upper layer of the liquid, when a deposit of copper will form on the metal if the mixture is unsafe.

Bordeaux mixture deteriorates very rapidly upon standing. It should, therefore, always be freshly made each time before use. If for any reason a quantity of spray is made up and cannot be used within a few hours it may be preserved in good condition for an indefinite time by adding a small amount of common sugar, one ounce for every eight pounds of stone lime or 10 lbs. of hydrated lime in the spray.

(ii) *Bordeaux mixture powders*—Various commercial companies have placed ready-prepared Bordeaux mixture powders on the market for immediate use by adding the requisite amount to water. Their use is less cumbersome and less time killing than the preparation of bordeaux from copper sulphate and lime. The companies furnish details about their preparation and the quantities to be used.

(iii) *Bordeaux with Resin*—Resin is added to Bordeaux mixture in order to increase the adhesiveness of Bordeaux and is needed for plants having a very smooth or glaucous surface and also when spraying is done during heavy rains. This may be prepared as follows:

Washing soda	1 lb.
Resin	2 lbs.
Water	1 gallon

Boil the water, then add the soda. When dissolved, add the resin and boil for about an hour stirring continually. Add this when cool to Bordeaux mixture prepared as usual at the rate of one gallon to every 24 gallons of bordeaux.

Bordeaux with resin has given good results in the spraying of areca palms for koleroga as the resin increases the adhesiveness to a marked degree. A weaker solution containing only half as much resin has also proved to be superior to ordinary bordeaux in spraying potatoes in the Khasi and Jainta Hills. In both the cases spraying has to be carried out during the period of heavy rain.

(iv) *Burgundy mixture*—This mixture may be substituted for Bordeaux mixture in localities or situations where it is impossible to obtain good lime. Its fungicidal value is considered about equal to Bordeaux mixture and it does not leave so much colour on the foliage or fruit as does the latter. It probably does not adhere so well through rains as Bordeaux.

Copper sulphate	10 lbs.
Washing soda	12½ lbs.
Water	50 gallons.

This is the standard mixture which is extensively used against potato blight in Ireland with excellent results.

Make up exactly as in Bordeaux mixture, using soda instead of lime. The materials used should be guaranteed 98 per cent. pure. Test as in the case of Bordeaux with blue litmus paper and add more soda solution if it reddens. Excess of soda may cause scorching, and can be detected by red litmus paper turning blue, in which case more copper solution must be added.

(v) *Ammoniacal Copper Carbonate*—This is a copper spray which leaves no visible deposit. In spraying fruits nearing maturity with Bordeaux an unsightly deposit is usually left on the fruit. In order to avoid this undesirable result this spray was devised. It is not usually as effective as Bordeaux but in some cases it may be desirable to use it in order to escape the bad effects of Bordeaux. The usual formula is as follows:

Copper carbonate	5 or 6 ounces
Ammonia	3 pints
Water	50 gallons

To prepare this spray first dissolve the copper carbonate in the ammonia and then add the water.

(vi) *Lime sulphur solutions*—Lime in combination with sulphur is used for spraying plants against fungus and insect attack. It is particularly effective against the powdery mildews. It can be prepared in homes or bought from commercial firms who make it and sell in barrel lots or smaller amounts. The prepared solution is diluted on a basis of a test of 32° Baume, which is the usual standard of concentration. For fungicidal purposes it is used at a strength of 1 gallon of solution in 25 to 30 gallons of water.

A home-prepared product may be prepared as follows:

Sulphur	1 to 1½ lbs.
Fresh stone lime	½ lb
Water	½ gallon.

The lime is slaked in the cooker. The sulphur and the water are thereafter added and boiled briskly till the sulphur is dissolved (about 45 minutes) stirring continuously and keeping the cooker covered. Water is being added as it boils down. When boiling is completed it is allowed to settle and the clear liquid is used.

Success in spraying

The success of a spraying done to control or prevent a particular disease will depend on:

(i) The selection and use of the proper fungicide. The crop to be protected and the parasite to be eliminated must be given consideration in making the choice of the fungicide since our crop plants show great variations in sensitiveness to injury from sprays and the different parasites exhibit various degrees of tolerance to poisons. For these reasons we would select for example, Bordeaux mixture for potato blight, rather than lime sulphur; lime sulphur or other sulphur fungicides for powdery mildews rather than Bordeaux mixture.

(ii) The application of the spray at the right time. The timing of applications and their number can be regulated only by a knowledge of the life-history of the parasite. For examples plants to be protected from a particular disease should be covered by a fungicide immediately before the time when infection is probable.

(iii) The thoroughness of application. That is the plants should be thoroughly sprayed and the surfaces to be protected, whether leaves, twigs or fruit completely covered by the fungicide. It is important that all parts of the plants to be protected should be covered by the spray, the topmost portions as well as the lower parts, the inner branches of the tree as well as the outer, both sides of the leaves and the entire surface of the fruits.

Spraying Apparatus

The selection of spraying apparatus is a subject upon which no extensive advice can be offered. It is best left entirely to the ingenuity of the plant grower. There are in general three kinds of pumps in common use. Bucket pumps are made for use with small amounts of the fluid in ordinary buckets. They are intended for small gardens but are not convenient for extensive sprayings.

Knapsack sprayers are equipped with straps so that they may be carried on the back and are so arranged that the operator pumps with one hand and holds the nozzle with the other. They are suited for more extensive work and are used for low shrubs or potatoes.

Compressed-air hand sprayers are small capacity sprayers comparable in use to the knapsack types. This type consists of cylindrical tank with the central pump, which compresses the air, thus forcing the spray out under pressure.

The barrel pumps are large pumps intended for attachments to barrels. They possess more general usefulness than either the knapsack or bucket pumps on account of the greater amount of fluid carried and the capacity for work.

II. Dusting

Fungicides that are applied dry in the form of powder are known as *dusts* and their application *dusting*. Compared with spraying dusting has certain advantages and disadvantages. They are:

Advantages of dusting—(i) Requires less time for application. Some claim it takes only 1/5th as much time as spraying. (ii) Requires no water. This is quite important an item when water is scarce or not near at hand. (iii) Labour cost of application is less. (iv) Difficulties of preparation are reduced; the dust may be purchased ready for application, the expense of time and labour for preparation and the cost of mixing vessels disappears. (v) Simpler and lighter apparatus for application is required. The initial cost of the dusting outfit is less than that of spray outfits of corresponding kinds.

Disadvantages of dusting—(i) Materials cost more per acre than sprays. (ii) In many cases dusts are less effective than sprays. (iii) Dusts are more difficult to apply satisfactorily in windy weather.

Materials for Dusting

Compared with spraying materials dusting materials are few. Any fungicide which is to be applied as a dust should satisfy two prime qualifications: *fineness* and *lightness*. Materials which satisfy these and are commonly used for dusting are mentioned below:

(i) *Sulphur*—Sulphur is most commonly and widely used. It is used in the finely divided form known as 'flowers of sulphur', chiefly against the powdery mildews. The toxicity of a sulphur dust is influenced by the size of the particles which should be at least fine enough to pass a 200-mesh sieve; many of the commercial brands are still finer. It is convenient for garden use, but is likely to be replaced in field practice by the lime sulphur combinations. It should be dusted early in the morning when the leaves are still damp from dew as it adheres better than if the leaves are dry. The best results are obtained on fine dry days while rain is injurious as the powder is readily washed off.

(ii) *Copper dusts*—Copper-lime dusts have been used as substitute for the copper-containing sprays like Bordeaux or Burgundy. These are composed of 5 to 25 per cent of a dehydrated or monohydrated copper sulphate with a dilutent, generally hydrated lime. In some trials the copper lime dusts have given as good control as spraying but the majority of comparative tests have shown them to be decidedly inferior.

Types of Dusters

For the application of dusts different outfits are necessary than those used for the application of liquid fungicides. Dusters are of numerous kinds and vary from the simple powder gun to the portable power dusters. 'These are designed to meet the same needs as the sprayers and include simple shakers, bellows dusters, powder guns or plunger dusters, and machines with fan blowers including small hand or crank dusters, sprocket power machines for attachment to a cart or light wagon, and special traction-power dusters for truck or field crops or vineyards with row attachments and gasoline-power dusters for use with either field or orchard crops'! The airplane has recently been used for the distribution of dusts.

When to apply Fungicides

Most fungicides are used to prevent infection by

covering the healthy part of the plant with a fungicidal layer. They are usually applied when the plant is in leaf, the first application at or just before the period when the disease appears, and others often enough to replace the fungicide when it gets washed off by rain, or begin to lose its effect or when new leaves are formed that require to be similarly protected. If rain falls before the fungicide has properly dried on the leaves, say within 12 to 24 hours, much may be lost and an early repetition may become necessary. Hence applications of fungicide should be done where possible in dry weather and as early in the day as practicable.

III. PAINTS

Various paints are used to protect tree wounds so that pathogenic fungi or bacteria do not gain entrance into the plant body through these wounds. The most commonly used ones are:

(i) *Bordeaux Paste*—One of the best dressings that has been devised to protect tree wounds against infection by wood-rotting fungi is Bordeaux paste. This contains the same ingredients as Bordeaux mixture but with much less water in proportion to copper and lime. The usual formula is:

Copper sulphate 1½ ℔s. in one gallon of water
Quicklime 3 ℔s. slaked in one gallon of water
Mix in equal parts.

This makes a paste which is applied to the wound in the same manner as white wash.

(ii) *Grafting Wax*—A useful grafting wax for covering tree wounds and top working is made either by melting equal parts of beeswax and resin or by using 4 parts of resin, 2 parts of beeswax and half part tallow. This is a very effective paint for keeping away fungus spores and bacteria.

(iii) *White Wash*—Its common use is for the covering of the trunk and large limbs to prevent sunburning, especially when the trees are severely pruned or when they may be considerably opened up for treatment of diseases.

A formula to prevent sunburning of trees is:

Hydrated lime	50 lbs.
Water	8 gallons.
Salt	4 lbs.

CHAPTER VI

SEED TREATMENT

In the majority of cases the parasite is carried either on or within the seed: (i) as spores lodged upon the surface or developed in the interior; (ii) as the vegetative body of the parasite, *mycelium, bacterial cells* or *nematodes,* either external or internal and (iii) as *sclerotia* or resting bodies on the surface or sometimes within vegetative reproductive structures as may be illustrated by the *Rhizoctonia* disease of potatoes.

The diseases that are transmitted through seeds are attacked in seed treatment. Enormous number of methods of seed treatments have been tried. They may be conveniently dealt under two heads: *chemical* and *physical.* Many of the methods have not found general adoption. Some of the features demanded to insure general adoption are:

(a) A reasonably low cost.

(b) No injury, or but slight injury to the seed when used at the strength required to kill the pathogene.

(c) An easy or relatively simple method of application.

I. Chemical Methods

The object here is to treat the *seed* with some chemical which will prove toxic to or prevent the development of the resting stage of the pest without injury to the germination or growth of the *seed.* The principal chemical compounds or mixtures which are in general use are as follows:

(i) *Copper carbonate*—Powdered copper carbonate is an effective seed disinfectant. It is available on the market in two strengths: the pure copper carbonate containing 50 to 54 per cent. metallic copper and the diluted mixtures containing 16 to 20 per cent. metallic copper. It is recommended for bunt of wheat and is known to have certain advantages: it causes no seed injury and wheat may be treated months before sowing without any decrease in effectiveness. It is not recommended for other cereals but has been used effectively for the control of damping off of miscellaneous garden vegetables. For wheat 2 to 3 ounces per bushel is effective; larger amounts may be used when the seeds are badly smutted.

(ii) *Corrosive sublimate*—It is the most effective disinfectant for seed potatoes, sweet potatoes and bulbs of flowering plants. Three slightly different formulas have been employed: (a) the cold or standard solution or steep, generally one ounce in 7½ gallons of water (1:1,000) or sometimes stronger, one ounce to 6 gallons; (b) the hot solution used at either standard or double strength for 5 minutes at 120° to 122°F., and (c) the acidulated mercuric chloride of standard strength plus one per cent. hydrochloric acid (1-10-1000) for one hour or for 5 minutes in a stronger solution (1-10-500). It can also be used at a strength of (1:500) for disinfecting tools when cutting out diseased parts of plants.

(iii) *Inorganic Mercury Compounds*—A number of inorganic mercuric compounds under different trade names have been placed on the market for seed treatment. Some of these contain the following besides mercuric chloride: sodium chloride (*Fusafine*), formaldehyde (*Sublimoform*), copper sulphate (*Weizenfusariol*), phenol and alcohol (*Betemal*). Some of the other inorganic mercury compounds used are: mercuric iodide (*Abavit* B), mercury rhodanide (*Urania*), and a mercury-copper-arsenic compound (*Tillantin* C). The directions for their use are furnished by the manufacturers along with the products. Some of these have been found very effective seed dressings.

(iv) *Organic Mercury Compounds*—Various preparations with the main active constituents in the form of organic mercury compounds combined with dilutents and stabilizers have been placed on the market some for use in liquid form, others as dusts. They have been marketed under various trade names, some of which are: *Chlorophol, Uspulun, Semesan, Semesan Bel, New Improved Semesan, Germisan, Ceresan, New Improved Ceresan, Agrosan.* The directions for use of these compounds are furnished by the manufacturers. These products have been found very effective materials for seed dressings against numerous diseases. They do not injure the seed in any way; it is even claimed that they increase the germinating powers of seeds in some cases.

(v) *Formaldehyde*—Formaldehyde is a gas and is available on the market in a 36 to 40 per cent. aqueous solution. It has been extensively used for the control of stinking smut of wheat, oat smuts and the covered smut of barley and for potato tubers, especially for the control of common scab. It has been used both hot and cold; the principal formulas being 1 pint to 30, 40, 50 or 60 gallons of water for the cold dips.

Methods of Chemical Disinfection

Seed may be subjected to the action of the chemical in a variety of ways. The more suitable and convenient methods are mentioned below:

(i) *Sack method*—This is adapted to treating either small quantities of seed in muslin or cheese-cloth containers or larger quantities in partly filled gunny bags. The seed is dipped for the requisite length of time in the fungicide of desired strength and drained and planted at once or, if necessary, given an approved after bath to reduce seed injury.

(ii) *Open-tank method*—The seed is emptied into an open barrel, tank or vat of appropriate size containing the treating solution. The seed is allowed to remain in the solution for the specified length of time after which the solution is drained off. One advantage of this method for cereals is that it makes it possible for the operator to skim off the bunt balls and the light seed.

(iii) *Sprinkling*—This method is mainly applicable to formaldehyde disinfection. The grain is sprinkled with the solution of requisite strength, shovelled over to distribute the moisture evenly; then piled and covered with a tarpaulin or sacks for 2 to 5 hours before drying.

(iv) *Dusting*—This method is used in applying the dry powdered materials. The fungicide in a very finely divided form is applied in such a manner as to coat the seed with a fine film of dust particles. Dusts may be applied to small quantities of seeds by shaking the dust and the seed together in dry glass jars. For fairly large quantities of seeds hand-operated dusting machines may be utilised.

II. Physical Methods

Heat should be used for the killing of the parasite only when no fungicidal treatments are available. Hot water treatment is effective for external seed-borne parasites but the method is so cumbersome that it is not employed when chemicals are available. It is mainly internal parasites which it is not possible to contact by chemicals without injuring the seeds which demand the use of heat as a therapeutant.

(i) *Disinfection by hot air*—Some of the seeds can be disinfected by hot air. It has been found that the internal mycelium of the late blight fungus in potato tubers can be killed by hot air treatments. A temperature of 40° to 49°C for 4 hours or 30°C for 65 hours is effective. But the method is too cumbersome to be adopted for general use. Effective elimination of anthracnose from cotton seed has been accomplished by a predrying at 122°F., for 36 hours or at 140°F., for 24 hours followed by an exposure of 8 hours at 205°F., with little or no decrease in germination.

(ii) *Disinfection by hot water*—Hot water treatment of seeds has been employed mainly for the treatment of wheat or barley for the loose smuts. The treatment comprises the following operations: (a) pre-soaking to initiate fungus growth, (b) pre-steeping in hot water a few degrees below the treating temperature, (c) immersion in the bath at the selected temperature and for the requisite time, (d) immersion in a cooling bath to prevent seed injury by too long action of the heat, and (e) drying of the seed by spreading out. The selected temperature for immersion varies from 118° to 129°F., for the different seeds affected with the different diseases and the time for treatments varies from 5 to 15 minutes.

CHAPTER VII

SOIL DISINFECTION

Many plant pathogenes, fungi, bacteria and nematodes spend a part of their life cycle in the soil and invade any suscept when opportunity offers. Many of the diseases caused by these soil-infesting organisms can be prevented by soil disinfection while others must be handled by other methods.

Soil disinfection as a means of disease control is rarely practical for extensive field plantings but can very successfully be employed for sterilizing soils of seedbeds or small plots.

Disinfection may be carried out by (i) *physical* and (ii) *chemical* methods.

I. Physical Methods

In the physical method heat is the principal agent employed and it may be furnished by oven baking, direct firing, hot water, flame thrower, some electric device or steam. In all cases of heat treatment the operation must be carried out previous to planting.

(i) *Direct firing*—This is a primitive method but is used even at present. With certain hill tribes this is a regular practice and they burn the soil before making any planting. It is carried out by burning straw, plant refuse or wood on the ground to be sterilized. It has been found sufficiently effective when burning is done thoroughly and carefully.

(ii) *Hot water*—Hot water may also be utilised for killing plant pathogenes in the soil. Fungi and nematodes in 4 inch pots have been effectively controlled by 5 minutes' submersion of the pots at 98°C.; in 8 inch pots they have been killed by drenching them with boiling water at the rate of 5 gallons per cubic foot.

(iii) *Steam*—Methods of sterilizing the soil with steam have been devised and developed in foreign countries but their applications are limited on account of prohibitive costs, and complications of operation.

(iv) *Flame thrower*—A flame thrower may be very satisfactorily used for sterilizing the seed bed soils or small plots. It gives a very strong flame the temperature of which is about 2000°C., and this flame can be utilised in burning the soil. The operation is easy and effective.

II. Chemical Methods

Many different chemical agents have been tried either as a drench or in powdered form but only a limited number have ever come into general use.

(i) *Formaldehyde solution*—This may be used as a drench for the disinfection of seedbeds. The amounts recommended have been one volume of the commercial product to 100 volumes of water applied at the rate of one gallon per square foot. The soil is prepared, the drench applied, the soil then covered with a heavy canvas for 24 to 48 hours and then allowed to air for one week to 10 days before seeding or planting.

(ii) *Formaldehyde dust*—This has been recommended for the control of damping off and other soil fungi according to the following method: Mix 6 ounces of commercial formaldehyde dust with each cubic foot of soil by shoveling it over several times. Place the soil in pots and plant the seeds at once if desired but water as soon as planted. If weak or low vitality seeds are to be used, they should not be planted until 24 hours after the treated soil has been placed in the pots.

(iii) *Mercury compounds*—Effective control of soil infesting fungi has been obtained by mercurous chloride, corrosive sublimate and organic mercury compounds such as *Uspulun, Semesan, Germisan,* etc. Mercuric chloride (1:1000) may be used as a soil drench at the rate of 2 gallons per square yard. Semesan may be used at the rate of one ounce to 3 gallons of water applied at the rate of 2 gallons per square yard.

(iv) *Kerol*—This has been found effective in controlling soil fungi. The soil is drenched by a solution containing one part of kerol in 500 parts of water.

BIBLIOGRAPHY

Bancroft, K.—A Handbook of the Fungus Diseases of West Indian Plants. 1910.

Bennett, F. T—Outlines of Fungi and Plant Diseases. 1924.

Brooks, F. T—Plant Diseases. 1928.

Butler, E. J—Fungi and Diseases in Plants. 1918.

Cook, M. T—Diseases of Tropical Plants.

Cunningham, G. H—Plant Protection by the Aid of Therapeutants. 1935.

Duggar, B. M—Fungus Diseases of Plants. 1909.

Eriksson, J.—Fungoid Diseases of Agricultural Plants. 1912.

Freeman, E. M—Minnesota Plant Diseases. 1905.

Harshberger, J. W—A Text-Book of Mycology and Plant Pathology. 1917.

Heald, F. D—Manual of Plant Diseases. 1932.

Hubert, E. E—An Outline of Forest Pathology. 1931.

Kuster, Ernst—Pathological Plant Anatomy. 1903.

Lodeman, E. G—The Spraying of Plants. 1909.

Martin, H—The Scientific Principles of Plant Protection. 1928.

Massee, George—Diseases of Cultivated Plants and Trees. 1915.

McCubbin, W. A—Fungi and Human Affairs. 1924.

Milburn, T. and Bessey, E. A—Fungoid Diseases of Farm and Garden Crops. 1915.

Owens. C. E—Principles of Plant Pathology. 1928.

Smith, E. F—Bacterial Diseases of Plants. 1920.

Sorauer, Lindau. Reh—Manual of Plant Diseases. 1922.

Stevens, F. L—Fungi which Cause Plant Diseases. 1913.

Stevens, F. L. and Hall, J. G—Diseases of Economic Plants. 1910.

Tubeuf, K. F. von and Smith, W. C.—Diseases of Plants Induced by Cryptogamic Parasites. 1897.

Ward, M.—Disease in Plants. 1901.

Weiss, F. E. Imms, A. D and Robinson, W.—Plants in Health and Disease. 1916.

Whetzel, H. H—An Outline of the History of Plant Pathology. 1918.